T0291268

Statistics for Public Policy

Statistics
for Public Policy

A Practical Guide to Being Mostly Right
(or at Least Respectably Wrong)

JEREMY G. WEBER

The University of Chicago Press
Chicago and London

The University of Chicago Press, Chicago 60637
The University of Chicago Press, Ltd., London
© 2024 by The University of Chicago
All rights reserved. No part of this book may be used or
reproduced in any manner whatsoever without written
permission, except in the case of brief quotations in critical
articles and reviews. For more information, contact the
University of Chicago Press, 1427 E. 60th St., Chicago, IL
60637.
Published 2024
Printed and bound by CPI Group (UK) Ltd, Croydon, CR0 4YY

33 32 31 30 29 28 27 26 25 24   1 2 3 4 5

ISBN-13: 978-0-226-82565-6 (cloth)
ISBN-13: 978-0-226-83075-9 (paper)
ISBN-13: 978-0-226-83074-2 (e-book)
DOI: https://doi.org/10.7208/chicago/9780226830742.001.0001

Library of Congress Cataloging-in-Publication Data

Names: Weber, Jeremy G., author.
Title: Statistics for public policy : a practical guide to being mostly
    right (or at least respectably wrong) / Jeremy G. Weber.
Description: Chicago : The University of Chicago Press, 2024. |
    Includes bibliographical references and index.
Identifiers: LCCN 2023020233 | ISBN 9780226825656 (cloth) |
    ISBN 9780226830759 (paperback) | ISBN 9780226830742
    (ebook)
Subjects: LCSH: Social sciences—Statistical methods.
Classification: LCC HA29 .W399 2024 | DDC 300.1/51—dc23/
    eng/20230530
LC record available at https://lccn.loc.gov/2023020233

♾ This paper meets the requirements of ANSI/NISO Z39.48-1992
(Permanence of Paper).

For Sara, who contributed to this book
in more ways than she knows

Contents

Preface

This book began with a practical need. When I began teaching quantitative methods after working in a federal statistical agency, I realized that students were leaving my classes without the confidence and ability to use what I had supposedly taught them. I was teaching about tools, not about the craft of using them to build things.

The textbooks for my classes precisely defined and cataloged statistical concepts and formulas, but fell short when illustrating their practical value. My policy-focused students and I needed more than examples where X is the number of blue balls and Y is the number of yellow balls. We needed a statistical mentor to bridge theory and practice—and not academic practice, but practice for the mayor's staff as they sized up issues, considered options, and made decisions.

Later, my experience at the White House Council of Economic Advisers also pointed to the need for practical statistical mentorship. Whether they were recent undergraduates or established academics, new staff usually needed several months or more to recalibrate and excel in competently applying the basics in a policy setting. For example, new staff (myself included) would initially report statistics and be unprepared for fundamental questions of interpretation, such as "Does the definition of income used to measure poverty include the value of public assistance such as housing vouchers and health insurance subsidies?"

Thus, my policy and teaching experience has pointed to the need for an engaging and colorful guide to using statistics in policy, which this book seeks to be. I hope it aids you in picking up the tools of statistics and using them well.

1

The Big Picture

Wielding Knowledge and Being Right

Statistics surround us, as do calls for decisions to be evidence based or data driven. But do you have the confidence and ability to use statistics in the room where people are setting policies that affect every student in the school or every senior citizen in the country?

We wield tools that we know well, not ones that we fear misusing and causing us problems and embarrassment. It is one thing to start a chainsaw and press the trigger a few times to make the chain spin. It is another thing to take the saw and use it to begin cutting the dead oak towering over the garage. By helping you master what is most important, this book builds your wielding knowledge, which is the knowledge that we hold firmly enough that we have confidence to use it when important things like health, money, and reputation are on the line.

Wielding knowledge includes knowledge that enables clear communication to the director, the lawmaker, or the public. They must get the point of our statistics. We don't have wielding knowledge if we cannot communicate the meaning and relevance of our statistics to our audience. Having knowledge to calculate or find the right statistics but not clearly communicate them is like knowing how to start the chainsaw but not how to cut down the tree.

Clear communication is more than creating a pleasing graph. It requires perceiving and explaining the meaning of the statistics for the situation at hand. Much statistical education focuses on making calculations and explaining their narrow meaning, such as explaining that the median value is the one where half of all observations have a greater value and half have a lower value. But having the median in hand and knowing its definition is only the first step toward using it to inform a decision, and it is usually as far as our statistical education takes us.

Understanding the relevance—or lack thereof—of a statistic in a particular context is outside the scope of the typical textbook or class. After her stats class, a student can calculate the lead concentration in the drinking water of the average sample home, the concentration at the 90th percentile, and the percentage of homes with concentrations over a certain threshold. But classes rarely equip us to discern which of the numbers matters the most or to argue that the city has a lead problem. This book equips you to do both.

This book complements textbooks that comprehensively cover statistical concepts and formulas. To them it adds principles for using statistics to further the understanding of people making decisions, principles that reflect my experiences in public policy settings that brought into high definition what we must firmly understand if we are to add clarity instead of confusion to policy discussions. These are essential principles that we must keep fresh in our mind if we are not to make a bumbling mistake and contribute to misunderstanding and bad decisions. Here the goal is a firm grasp of matters of primary importance, ones for which being wrong can make everything else wrong or irrelevant.

Mastering issues of primary importance is key to being right. By "right," I mean presenting work that would not embarrass you if a statistically savvy journalist sifted through every piece of it from start to finish. It is free of mistakes, and its reasoning can be defended before knowledgeable people. People can still question your work and even argue for making other judgments or using other methods. And your statistics and predictions can be far from the mark in the end. But right work is not work that is unassailable and shows perfect foresight. It is work that has a good shot of being roughly right given the information and the constraints of the moment. It is work that is respectable in all its details even if it turns out to be wrong in hindsight.

For the Policy Aide and You

This book is written with the policy aide in mind, one who uses her statistical savvy and insight to support a policy maker. The aide occupies a middle space bordered on one side by academics who speak to other academics and on the other side by lobbyists and activists who seek the surest way to advance a particular policy. The policy maker benefits from an aide who understands data and approaches them evenhandedly, including in areas where she has strong policy convictions. Her work may encompass a full-blown analysis of a data set or just the thoughtful interpretation of someone else's statistic.[1]

The aide does not presume to be the policy maker but is also not a data lackey, unearthing whatever feel-good numbers that the boss wants. She is akin to a lawyer serving a client. The client is best served by a lawyer who provides a frank and trustworthy assessment of the evidence and the law, not one who leans upon weak evidence and contorted legal interpretations that will not stand the scrutiny of jury or judge. For the good of the client, the lawyer must do principled, independent thinking.

With the aide in view, the thrust of this book is practical and client focused. Statistical training often focuses not on the aide who serves a boss or client but on the researcher who writes to researchers about research questions. Researchers participate in perpetual academic conversations about the underlying nature of things across time and place. The conversations usually value the most generalizable knowledge. The experience of one school district or city is interesting only so far as it contributes to knowledge about schools and cities in general. The researcher asks: "Do school resources improve student outcomes?" The aide is asked—and seeks to answer—"Should Pittsburgh Public Schools borrow money to install air conditioning in its buildings?"

The aide may draw from research on the broad question of school resources and student outcomes, but the bulk of her work will involve the particulars of the case and the role of the policy makers she supports. In addition to prioritizing generalizable knowledge, academic discussions rarely produce conclusions firm enough to point to a decision. The evidence may hint at a way forward, but more research is always needed. The decision is perpetually postponed, awaiting more and better data. The aide, though, must work with the data in hand because the school board meets next week to vote on the borrowing proposal.

By addressing the aide, this book serves those addressing a diverse and practically minded audience. Whether your field is public health, education, journalism, administration, or policy, the examples and insights will help you cultivate the statistical savviness that organizations and governments need and expect from people in management. Our data-rich world has great demand for people who can wrap their minds around data and interpret summaries of them. The demand is for people who can bring statistics into discussions in ways that help people better understand an issue instead of distracting or confusing them.

This book should help the beginner and the advanced. It is not an introduction to statistics or an advanced dive into statistical subfields; it is accessible to the beginner and stimulating for the advanced.

It will help those with basic statistical training better understand what they have just calculated or read. Statistics textbooks calmly describe the formulas for standard errors and p-values and more over several hundred pages. The tone and design have their purpose, but they make it difficult for the beginner to identify what statistical concepts are most important for practical decision-making or what pitfalls one can encounter along the way. With its emphatic voice and practical focus, this book will make you more excited to read your textbook carefully by showing the practical relevance of its content.

There is also much here for the advanced, the PhD who wants to better communicate statistics to policy makers, a skill rarely taught in statistical education. For example, good communication involves adjectives—and reasons for them—that put a statistic in perspective for the audience. P-values and statements about statistical significance won't do. Read on for guidance on assessing magnitude. Part of improving communication is also improving our understanding of the basics, which may have become nebulous to the PhD who graduated a decade ago. This book will restore some of the understanding that often erodes with time as experts speak to themselves and not to people who have never heard the terms or thought of the concepts.

If I ever understood the difference between a standard deviation and a standard error, I lost it between graduate school and the first time I taught Introduction to Quantitative Methods. As sure as bread becomes stale, it is easy to pursue complex methods while becoming estranged from the basics. We remember the words—the standard error remains part of our vocabulary—but we slowly lose hold of the concepts behind the terms until we cannot explain them to someone who never understood them. With its practical illustrations, this book can serve as a refresher that revives our understanding of the standard error or the correlation coefficient.

In short, this book is probably for you.

To Statistical Skeptics: Statistical Claims Are with Us Always

A statistic is a summary of observations, which might be cars using a road, people becoming sick, or crimes being committed. As useful as summaries may appear, many are skeptical of statistics, arguing that they can be used to prove any point and support any agenda. You may be skeptical of statistics or just indifferent to them. Statistical claims, however, are with us always, with or without statistics.

Imagine that your country is experimenting with new agricultural policies. The central government has prohibited food imports and has specified new laws governing what farmers can plant and the price at which they can sell their crops. The policies fail and bring famine to the land. Anecdotes of people starving are everywhere. The leadership in the central government, however, claims that there is no problem. They state that any reports of poor harvests and malnutrition are isolated occurrences.

Alternatively, imagine that you work for a nonprofit organization that intervenes in cases of domestic violence and works to rehabilitate the perpetrators. After working in the organization for a year, your observations lead you to question whether the intervention helps or whether it even enables violence. The organization's executive says that donors find the intervention approach appealing and are pleased by the success stories highlighted in periodic reports. He sees no need for a systematic assessment of the organization's work.

These two examples illustrate that a world without statistics is not a world without statistical claims. In the agricultural example, political leaders claimed that there was no famine and that harvests were plentiful, a statistical claim lacking only in statistical evidence. In the nonprofit example, the organization shared success stories with donors, implicitly claiming that such stories are common and perhaps the norm. Again, a statistical claim lacking only statistics. We regularly make or absorb claims about what is common or rare in the world around us. Our belief vacuums fill quickly, with or without statistics to support them.

Discussions about public problems or programs inevitably involve claims about what is rare and inconsequential or pervasive and serious. Is there a problem that warrants a policy change or a new program? Does the program have its intended effect? If we lose interest in verifying whether claims have a factual basis, we are cut loose to assert claims that fit our desires but not our reality. In the agricultural example, the desire of political leaders to retain power motivated them to claim that their agricultural policies had succeeded. In the nonprofit example, the desire to maintain funding spurred the executive to present cherry-picked examples as representative of the organization's work.

Without statistics to describe the world around us, claims made in policy discussions rest on personal experience and other anecdotal evidence. The political signs in the yards around my neighborhood become reflections of national political preferences. My neighbor's cousin's wife who had a bad reaction to a vaccine becomes an indication of the risk of the vaccine and perhaps all vaccines. My brother's neighbor who sells his food-assistance benefits for cash to buy cigarettes and beer becomes the archetypal assistance recipient. Soon we are confident about the right direction of policy on the basis of one or two observations.

Anecdotal evidence can be informative, and even largely representative, but two people can have contrasting observations that lead to contrasting conclusions. We might have two different neighbors who, for example, use their food-assistance benefits differently. And there lies the power of statistics: they provide us with summaries of many neighbors. When one person says that the benefits just finance cigarette and beer consumption and another says that they rescue millions from hunger, statistics say that more than half of the benefits go toward increasing the consumption of eligible food purchases in the month they are received.[2]

And so to the statistical skeptic, I encourage you to press on with statistics. Statistical claims will be with us always, so let's tether them to statistics.

To Statistical Enthusiasts: Statistics Is Like Making Maps, Not Selecting Destinations

After working in the White House Council of Economic Advisers for a year and a half, several people asked me whether data or statistics had mattered much for policy decisions. My response: "It depended." My experience on Pennsylvania Avenue is similar to what you will find on Main Street and everywhere in between.

Many of us statistical enthusiasts, ever ready to jump into a data-driven car, would benefit from reflecting on the limited (though important) role that statistics actually play in policy decisions. With so much data a click away, we often look to data for a pathway through canyons of political impasse and into the plains of good policy. We expect much from data and therefore say that our policy is evidence based, science based, or data driven. We let the evidence, the science, and the data identify the problem and its solution.

The notion seems reasonable, and it is reasonable to make decisions informed by systematic observation. Decisions based on whims, hunches, or lies may lead to solutions for phantom problems or phantom solutions for urgent problems. Yet we can err by asking from data what they cannot give.

Statistics as Mapmaking

It is easy to think that the link between data and decision is straightforward. But statistics cannot tell us what to do any more than a map can tell us where to go.

We do not ask maps to give us a destination. Rather, we put in a destination, and the map app gives various routes and their travel times. The map does not select the destination; it aids in choosing the best route to the user-provided destination. It can of course help us select among competing destinations, providing information like travel times to various destinations and the tolls associated with various routes. But the map does not tell us how to weigh travel time and toll cost, and much less how to weigh differences between the destinations themselves. And so it is with statistics and decisions about eligibility requirements for food assistance or discipline practices in schools.

The map analogy helps to illustrate another point. Maps are meant to accurately describe the actual relationships between one location and another and what lies between. In this sense, maps are descriptive, and mapmaking is a positive activity that states what is, not what ought to be. And yet in another sense, mapmaking involves value-laden decisions. The mapmaker must choose which features to include because only a few can be included. A mapmaker who attempts to show all features of a road—its speed limit, its grade, its dimensions, its material type—will bewilder the map user, and the map will fail at its most basic purpose. The mapmaker must also determine the orientation of the map. Will Antarctica be on top? Will the United States be in the center or to the side?

In the same way, producers and presenters of statistics choose which data are worth collecting and which statistics are worth calculating and presenting. We also choose how to present the statistics so as to prompt particular thoughts in the minds of our audience. We might say "More than 10 percent of the country's residents live in poverty" instead of "More than 85 percent are poverty-free." We might say that the program increased employment by only 1 percent or (equivalently) that it created more than 100,000 jobs. Such choices about framing can have large effects on an audience's thinking and decisions (Daniel Kahneman received the Nobel Prize in Economic Sciences for showing such effects).

If we should think of statistical analysis and its presentation as mapmaking, how should we think about the role of maps in policy decisions?

Statistics in Policy: From Decisive to Irrelevant

When and how do statistics matter for policy? Again, the map analogy is helpful. Sometimes there is clarity on the criteria for selecting a destination, in which case the information that the map provides is decisive for the decision: we're hungry, and we just want to find the closest restaurant, regardless of what it serves. Other times, there are multiple criteria, perhaps of varying importance to different people, and the map informs the selection of the destination but is not decisive: we're hungry, but some people prefer to drive a little farther to eat at a locally owned restaurant instead of a chain restaurant. Last, the information provided by the map can be irrelevant: some people have taken a vow to only eat at locally owned restaurants as a matter of principle and consider travel time irrelevant.

In summary, statistics can be decisive for a decision, informative but not decisive, or irrelevant. I'll consider each case in detail.

DECISIVE: Occasionally the map (statistics) is decisive for determining where we want to go. The number crunchers often take the wheel in cases of *Should I Care?*, *Which Method?*, and *Don't Go There*.

Should I Care? In *Should I Care?* cases, statistics determine whether an issue is a problem meriting attention. A news story can bring attention to what seems to be a widespread problem: three people interviewed said that they had been arriving late to work because city buses were running late. Alarmed, the leaders of the transit authority meet, thinking that a systematic problem in the bus service has emerged. They summon the data analyst to run statistics on the discrepancy between actual arrival times and scheduled times. The analyst returns with a striking finding: the discrepancy for the past month was the lowest on record. Closer investigation reveals that the news report reflected one route on which a driver had two bad days in a row for personal reasons. Upon learning this, the transit authority leaders quickly turn their attention to one of the ten other matters demanding their attention.

Which Method? Statistics can also be decisive in *Which Method?* cases—in these, the policy goal is clear and agreed upon but there is uncertainty as to which method best achieves it. Suppose that your agency runs a program designed to keep people convicted of domestic violence from repeating their crime. The program is having limited success because most participants drop out after a few months. The program can be run by meeting once a week for six months or twice a week for three months. Data showing that participants of the accelerated program have a 25 percent higher completion rate provide clear guidance on which format to choose (assuming that there were no other differences or effects of the formats). The data have spoken, and the better method is clear. End of discussion.

Don't Go There. Statistics can also be a showstopper in *Don't Go There* cases. This is where a statistic reveals that the policy has an unacceptable consequence or risk. Imagine that the White House finds that the downside of a policy proposal would fall entirely upon swing voters in a swing state. Whether in the White House, at city hall, or in the boardroom, it is easy to imagine outcomes that carry high political or public relations risk even though the policy behind them may benefit the average person. Such cases are not limited to political or public relations risks. They can involve other unacceptable outcomes. A social impact investing group may drop funding for an innovative battery when it learns that one in a thousand batteries inexplicably bursts into flames, a fatal flaw for an otherwise promising technology.

INFORMATIVE BUT NOT DECISIVE: More common are cases when the map is relevant but not decisive for deciding where to go. Most policy decisions involve weighing trade-offs—how happy are we with more of this and less of that? Statistics, though, provide no easy answer to trade-off questions. Suppose that a study showed that one suspension in the ninth grade increases the likelihood that the suspended student eventually drops out of high school. Another study indicates that suspensions improve the learning of students in the same class as the suspended student.

Do the findings of the two studies tell you, a school board member or superintendent, what to do? No. They highlight the trade-offs of keeping disruptive students in the classroom relative to removing them. The severity of the trade-off depends on the magnitude of the estimated effects. Does a suspension increase the likelihood of the suspended student dropping out by 5 percentage points or 50 percentage points (so is it increasing the probability of dropping out from the baseline probability of 5 percent to 10 percent or from 5 percent to

55 percent, making it more likely than not that the student drops out)? And does keeping the disruptive student in the classroom reduce the test scores of peers by 1 percent or by 20 percent?

Assessing the trade-off involves beliefs about the importance of the effects being traded off. How acceptable is it that the suspended student is harmed while other students benefit? The answer depends on beliefs about justice. One might argue that the student being disciplined is receiving the due penalty for her disruptions. The time outside the classroom and the lower likelihood of graduating are consequences of her choices, and they are proportional to the seriousness of the offense. Alternatively, one may reject the idea that the student should receive a penalty for disruptions, especially when it may stem from a difficult home life or other aspects of the environment. True justice demands restoration, not penalty. Any negative effects on the suspended student are inexcusable.

So, finding that suspension increases dropout rates is not decisive for a policy decision. Even a finding that suspension had no effect on dropout rates might not be decisive for continuing the practice. Suspensions might affect other outcomes, such as grades and likelihood of college admission. And even if the most credible studies show that suspension has no causal effect on scholastic outcomes, parents or school administrators may reject the policy because it violates convictions about how schools should treat students. Critics might say that punitive discipline like suspension should be abandoned in favor of restorative and inclusive measures. This leads to the third case, in which statistics are irrelevant for policy decisions.

IRRELEVANT: Last, the map can be irrelevant if we know that we do not want to go certain places. Some policies can be dismissed outright because they violate higher-order convictions. Consider the following fictional case:

> Many studies show that the children of poor parents also tend to be poor as adults. Learning this, a leader in a country with an authoritarian government enacts a forced sterilization program with the goal of reducing the poverty rate. Any poor man or woman of reproductive age

is forced to be sterilized. A careful study of the policy later estimates that it reduced the country's poverty rate by 25 percent over a decade. The estimates were convincing and robust.

A data-driven policy adviser in a democratic country learns about the study's findings and recommends the policy to a top government official, citing its effectiveness in reducing poverty. The government official responds by saying, "Forced sterilizations could eliminate poverty entirely, and we still would not do them."

The policy violated the convictions of the official or the people he represents such that no degree of effectiveness in poverty reduction would justify its adoption. Perhaps the official believes that the government has no authority to coerce people not to have children. Whatever the nature of the conviction, the findings of the study were irrelevant. The policy was a nonstarter.

The policy adviser may think that his recommendation is data driven and the official's response value driven and therefore subjective and perhaps even emotional. A recommendation of forced sterilizations, however, rests upon its own value judgments. For one, it assumes that income determines which lives are worth living, with lives below the poverty threshold not worth living. Under this assumption, the forgone lives are not a downside of the policy but an upside, because they are not worth living.

Beliefs about what is acceptable vary, and so will beliefs about the relevance of a statistic. Thus, one person may find statistics on the long-term effect of abortion on crime relevant, and perhaps even decisive, for abortion policy. Finding that crime went down would justify more access to abortion; finding that crime was unaffected could mean that those aborted would have become productive members of society, justifying limits on abortion. In contrast, those with strong convictions about the choice of a woman or life in the womb find such studies irrelevant.

Whether a statistic is decisive, informative, or irrelevant for policy can therefore depend on convictions about what is good and just. And these convictions can vary from person to person. This has implications for how we present statistics and discuss their implications. Statistical enthusiasts would do well to be patient with those who find our statistics less relevant than we do while also being mindful of our own assumptions and beliefs that link a statistic to a policy decision.

Two Themes

Two themes run through the book. Naming and explaining them at the outset will help you spot them as they surface along the way and see how they shape the use of statistics in policy.

The Statistician and Detective Must Work Together

One theme of this book is the importance of integrating statistical knowledge with knowledge of the particulars of a context and issue, which is gained with the skills of a detective. When a property suffers vandalism, the police department sends a detective to investigate, not a statistician. The detective specializes in specifics, the clues particular to the case. He does not seek to understand the probability that the average building will be vandalized or the growth trend in vandalism over the past few years. He seeks to understand one case and understand it so deeply that he can pinpoint who vandalized the Main Street Department Store on Thursday, December 15. The statistician's task is different. She looks for patterns revealed by hundreds or thousands of vandalism cases.

To inform a policy decision in a particular time and place, the policy aide must blend statistical and detective skills. Her detective side absorbs the details of the place and time and issue at hand—this city at this moment dealing with this particular rise in vandalism. Her statistical side steps back and looks at patterns in the city's data over time and even the data from other cities. She does not draw conclusions based on a few high-profile cases. The Main Street Department Store case might be well known among city council members but it might have specific circumstances that led to it. Most vandalism cases might have the mark of local youth being unruly, whereas the Main Street case suggests a more sophisticated effort to inflict serious damage. The statistician side of the aide helps her look across all the vandalism cases in the city in recent years so that the discussion does not assume that the Main Street case is the norm.

Integrating statistical knowledge with knowledge of the particulars of a context and issue yields statistics helpful to the policy maker. These statistics answer questions that point to causes and solutions, to questions like, Which types of vandalism are driving the overall increase? When exactly did the trend emerge, and what changed around that time? The answers to these questions are the fruit of integrating the work of the detective and the statistician.

Doing detective work on the particulars of a context and issue fits with the framework of Bayesian statistics, which begins with defining prior beliefs about a variable's distribution and proceeds with an updating of beliefs with new data. Although the limited attention given to uncertainty in this book does not use a Bayesian approach, several themes are at home with Bayesian statistics, including the incorporation of prior particular knowledge, the need to make decisions based on small and imperfect samples, and the updating of beliefs as better data become available.

Academic Standards May Not Apply

Another theme of this book (particularly in chapters 2, 5, and 6) is that statistical standards taught and used in academic settings may not apply in policy settings. The policy aide is not in an academic conversation and should not assume that academic standards are appropriate. Most statistics classes are taught by people who perform statistical analysis for academic articles in academic journals that are read by academic audiences. Their analysis must meet the standards used by journal editors and reviewers.

This includes standards for the representativeness of a sample—was it drawn in a way to make it reflective of a broader population? Standards regarding the uncertainty around an estimate—is the difference across two groups statistically different from 0 at the 95 percent confidence level? And standards regarding the credibility that an estimated effect is the real effect of a particular variable apart from other factors—was the 20 percent drop in crime caused by the change in policing practices or other things changing through time, such as a rise in job opportunities?

Naturally, the statistics professor teaches students the standards he applies in his academic work. Yet many of his students will never do statistics for academic journals. Their context and audience will be different. (Moreover, academic standards themselves are a complex and ever-evolving mix, varying across disciplines and within a discipline over time.) At times academic standards may make sense. The policy maker's tolerance for being wrong might match the low tolerance reflected in academic standards. People approving a new drug for the public will want great confidence that it has no serious side effects that could outweigh its benefits.

In other instances, applying academic standards may make little sense. Academics and policy makers such as elected officials often have a different aversion to being wrong. The academic wants to be nearly sure before concluding that a policy has worked. The forward-looking elected official might ask only that success is more likely than not. Similarly, the academic demands much evidence before concluding that the city has a problem with lead in its water, but the head of the city water authority springs into action if a problem is only mildly plausible. To use a statistical term, the academic and the policy maker have a different loss function, which assigns values to different types of errors.

Academic standards regarding the quality of evidence may also be inappropriate. The aide's analysis of the causal effects of police reform from five years ago might add tremendous insight to the city council's debate even though it lacks the rigor required by academic reviewers. Thousands of consequential policy decisions are made every day across diverse organizations, agencies, and levels of government. I speculate that many are made with little statistical analysis behind them. When the baseline level of evidence is a mix of intuition and anecdotes, it makes little sense to admit additional evidence only if it meets academic standards.

Whether particular data and statistics should enter the policy discussion depends on the specifics of the case, especially the extent of existing knowledge. As an example, statistics from imperfect, non-random samples were valuable when nearly nothing was known about the lethality of COVID-19, which chapter 2 will show. The same statistics could be worthless years after widespread data collection and hundreds of rigorous studies. More generally, academic research often lags policy decisions by years. Issues arise quickly, with no off-the-shelf academic analysis that is directly relevant. If such analysis exists, it probably uses data from years ago. Simple statistics using more current and relevant data might be more helpful even though the analysis may not meet academic standards.

Eugene Bardach and Eric Patashnik put it well in their book *A Practical Guide for Policy Analysis*. What they say about models applies to statistics. Their warning and recommendation is to "be careful . . . to avoid using the social scientific standard of adequacy for judgements about the realism of a model. . . . In policy analysis, the looser, but more appropriate, standard should be whether reliance on a model can lead to better results and avoid worse results than less disciplined guesswork."[3]

Appropriately deviating from academic standards, however, requires a firm grasp of statistical concepts and careful thinking about the particulars of a case. Read on to grow in these ways.

2

Know Your Sample and Data

Statistical endeavors begin before collecting data, but this part of the book meets you as you embark on learning from data in hand. You may have painstakingly collected the exact data you wanted or just stumbled across existing data that might be useful. Now it is time to know the sample and the data you have for it.

Preparing to Listen and Learn

Knowing your sample and data allows you to judge their relevance and quality. Relevance and quality in turn determine whether we should listen to what the sample data say.

In their textbook, Moore, McCabe, and Craig describe statistics as the science of learning from data.[1] Learning from data is harder than it sounds. We data users are human. We have beliefs about how the world works or should work and often face subtle incentives to establish or confirm a particular finding. Spend enough time working with statistics, and you will face the temptation to not learn from the data. When the data and the resulting statistics run contrary to our expectations or hopes, we will be tempted to cap our ears to what they are saying and turn to other data or anecdotes. Resist. We must humble ourselves before the data, reflecting upon them and submitting to them.

Learning from data is possible only if we have made a commitment before setting out to listen to them on the basis of their quality, not on whether we like what they say. We can know if we are capable of learning from the data only in those instances when the data say something surprising and challenging, and despite that we don't cap our ears. When the data confirm what we expected or wanted to be the case, we might be listening to them only because they say what our ears want to hear.

And yet it is also unwise to listen unquestionably to whatever statistics emerge from our first foray into the data, as if they carried infallibility and finality with them. A statistic that runs contrary to your expectation should motivate further exploration of the data in light of your purpose. Further exploring the data in response to an unexpected finding is not refusing to listen to the data. Refusing to listen is when you discard the data because of what they say, irrespective of their quality or relevance.

Know Your Sample: Origins, Purpose, and Generalizability

"Is This a Random Sample?" Is the Wrong Question

Every sample has a genesis story. Knowing it helps you assess the sample's suitability for the purpose at hand and for use in interpreting the resulting statistics. Statistics education often emphasizes the need for a randomly selected sample apart from any discussion of context or purpose. The privilege given to the generic random sample is so strong that at least one statistics textbook that I have used declares nonrandom polls and surveys to be scientifically useless. The assertion and ones like it have misguided the thinking of multitudes. Putting aside what "scientifically useless" might mean, any sample—even those randomly drawn—has limitations. Moreover, asking, "Is this a random sample?" is the wrong question, at least to start with. Before discussing the right questions, let's clarify key terms.

A random sample of houses, for example, is one in which each house in the population has the same probability of selection. The method of selecting houses and the pool from which they are selected do not favor colonial-style houses to ranch houses, old houses to new houses, or houses near the park to houses far away from it. As such, the percentage of particular types of houses in the sample should, on average, match the percentage in the population. So if houses near a park represent 10 percent of the population, we expect 10 percent of sample homes to be near a park. Any one sample's percentage of homes near the park might differ from the population percentage. But if we took many random samples and averaged the percentage of homes near the park across all samples (summing the percentages across samples and dividing by the number of samples), the average sample percentage should be near the population percentage. If, however, we had selected houses from a pool of people who had donated money for parks, we might expect the percentage of sample homes near the park to be much higher than in the population.

Two key concepts in that explanation are *representativeness* and *population*. We often want our sample to be representative of the population. But what is *the* population? All houses in the world? In the United States? In a particular county? It can be any of these. Your population is the group you want to say something about. If you are making estimates for the county planning board, your population is probably all houses in the county, in which case it is irrelevant that the average house in your sample is fifty years older than the average house in the state. Your focus is on the county, so you care that your sample average house looks like the county's average house.

Random sampling serves a particular purpose. A randomly selected sample should be representative of the population on average and therefore give unbiased estimates of population values. (An unbiased method gives us the right answer, the true population value, on average.) If our purpose is to credibly estimate the average age of houses in the population, having a sample of houses drawn only from older neighborhoods will not help us. For a credible estimate, we need a sample drawn from all types of neighborhoods.

As the county housing example illustrates, representativeness and population are context- and purpose-specific concepts. More broadly, diverse types of samples can provide information useful for policy decisions. Whether a sample is useful depends on context and purpose, which is why "Is this a random sample?" is the wrong question to start with. It often leads to a dead end and the discarding of data useful to policy makers. We must first ask other, broader questions. A real-world example will give color to the idea.

The *Diamond Princess* cruise ship set sail from Hong Kong in late January 2020 with 3,711 passengers on board. On February 1, a passenger tested positive for COVID-19, and subsequent testing confirmed more than seven hundred cases among those on board.[2] Few areas outside of China had many confirmed cases at the time, and policy makers across the globe were considering restrictions on international travel.

Did the passengers on the *Diamond Princess* represent a random sample? Certainly not. With an average age of fifty-eight, the sample was not representative of the global population or the population of high-income countries.[3] It may not even have been representative of the population of cruise-ship passengers. After all, the sample came from one particular ship that may have attracted a particular group of passengers.

Yet data from the *Diamond Princess* dramatically refined what we knew about the virus's infectiousness, the percentage of infections that produce symptoms, and the percentage of infections that would be fatal. For example, in early March 2020, the director of the World Health Organization reported that "evidence from China is that only 1 percent of reported cases do not have symptoms, and most of those cases develop symptoms within 2 days."[4] The statistic implied that nearly all infected people would know that they are infected and could quarantine, increasing the plausibility of the virus being contained. But in the case of the *Diamond Princess*, a little more than half of those who tested positive showed no symptoms.[5] A later review of more than two hundred studies on infection rates put the asymptomatic rate at 24 percent, less than the *Diamond Princess* number but much higher than the 1 percent reported by the World Health Organization.[6]

Despite being a nonrandom sample, the *Diamond Princess* passengers taught the world much about the virus at a time when we knew very little. It and similar imperfect samples provided important statistics for memos that I and colleagues at the Council of Economic Advisers were writing in February and March 2020. To use the language of the Bayesian statistical framework, we had little prior knowledge about the asymptomatic rate and the range of values it was likely to take. In such an uninformed and data-scarce moment, even small and imperfect samples greatly improved upon prior beliefs.

The example illustrates why the policy aide must not reflexively dismiss nonrandom samples: policy making barrels forward with the information at hand. One may choose to ignore existing data, but one cannot avoid making a decision, either about the policy or about giving credence to the data at hand, however imperfect they may be. The academic can postpone a firm conclusion about the benefits and costs of a government program, but the policy maker watching over that program has the ability to change some or all of it at any moment. Thus, the policy maker constantly faces making implicit decisions even if the status quo persists.

That said, there is good reason for the typical textbook's appreciation for random samples. Casual use of nonrandom samples can lead to broader or more forceful conclusions than what the sample warrants. We cannot use interviews with people visiting gun stores to gauge national public opinion on gun control policy. In some cases, having a randomly drawn sample may be paramount given our purpose and the need to generalize beyond the sample. The mean number of cars owned per household from a nonrandom sample might be wildly different from the population mean value. By extension, any estimate of how much a $1,000 per-car tax would cost the average household could grossly under- or overstate the true average burden.

The question of randomness, however, is too narrow a starting point to accommodate the diverse situations in which the policy aide might find herself. We must ask other questions of our sample that illuminate its strengths and limitations and guard us from tossing policy-relevant data.

Good Questions for the Sample

Three questions can reveal much about a sample's strengths and limitations. The questions concern origins—how did people enter the sample, and who are they? They concern purpose—what do I want to learn from the sample? And they concern generalizability—what can sample statistics say about people outside the sample? I use people as the unit of observation, but people could be replaced with classrooms or countries, subways or streams.

View the three questions as connected in a web rather than as a sequence in which each question is answered definitively before answering the next one. The questions of origin and purpose are tightly connected. Purpose can shape origins, determining who is sampled and which data are collected from them. It is also common for existing data to shape the purpose of research. This is especially true where time and resources limit the analysis to data already in hand. Also, origins and purpose affect generalizability. For example, if the data come from a regulatory form that only large farms must complete, generalizing the statistics to all farms might be a stretch. But if the purpose is to say something about farm practices on most cropland—and large farms manage nearly all cropland—the sample could work well.

ORIGINS: HOW DOES SOMEONE ENTER THE SAMPLE, AND WHO ARE THEY AS A GROUP?

Knowing how people enter the sample allows you to understand who they are as a group. How someone enters the sample may seem obvious if you did the sampling and conducted the survey. But people do not enter the sample just by being sampled. They enter by being sampled *and* by answering the survey. You may have randomly sampled businesses from a comprehensive list of small and larger businesses, but what ended up in the sample for which you have data are large businesses with staff dedicated to responding to outside information requests.

Understanding how someone enters the sample often requires understanding the details of definitions used to create the sample and collect the data. An example from US farm data illustrates that seemingly small details in definitions can have far-reaching implications for the interpretation of sample statistics. You might discover that you are a farmer according to the US Department of Agriculture!

The median farm household in the United States lost more than $1,000 from farming in 2018 and in 2020.[7] One might conclude that the current plight of the US farmer is on par with the Great Depression, and federal support of farms should be expanded. But appearances can mislead. The US Department of Agriculture's definition of a farm is "any place from which $1,000 or more of agricultural products were produced and sold, or normally would have been sold, during the year."[8] You might be a farmer! If I planted several apple trees in my backyard, my house and the postage-sized green strip around it might qualify as a farm. Consider the implications of the definition. If I planted apple trees, it would be for pleasure, not profit. And if I could include my property taxes as a farm expense, I would lose money from farming, even if I sold all my apples to upscale restaurants at high prices. My lack of profit from farming, therefore, does not reflect financial hardship or a fundamental problem with agricultural markets or farm policy.

The capacious definition of a farm would not affect median values if the number of so-called farms between, say, $10,000 in sales and the potential for $1,000 in sales was small. But it is not. In fact, one analysis suggests that simply removing the word *potential* from the definition of a farm reduces the farm population by about 20 percent.[9] The change in the population would of course also change the composition of any sample drawn from it and the resulting statistics.

How people enter and who ends up entering are especially important questions for what I call by-product samples. By-product samples come about as people use an app, swipe a credit card, visit the doctor, apply for a student loan, and so on. The data come from the churn of daily life, not from a research-driven, data-collection effort like a survey with a defined sampling strategy. By-product samples have become common as electronic devices have seeped deeper into our lives. They emerge as we take the temperature of our sick children, search for the travel time to Grandma's house, and swipe our credit card at the grocery store. Examples abound:

- Temperature data from smart thermometers
- Mobility data based on cell phone locations or use of map apps
- Agricultural yield data from software on tractors
- Social network data from social media
- Health insurance claim data from insurance companies or Medicare

These examples share the feature of only including users of a particular technology or service—users of smart thermometers, smartphones, precision agriculture, social media, and health insurance. The groups are large but probably distinct in some ways. Smart thermometer buyers might be fairly young, high-income parents who mostly use the thermometers on their young children. The resulting temperature readings may therefore reflect the health of a particular demographic. Similarly, spatially precise agricultural yield data may reflect only the fields of large farms whose planting and harvesting equipment have GPS-guided systems.

By-product samples can be very large and may have value despite their distinctiveness. At a minimum, a sample represents itself. Apple's mobility data, for example, are from Apple Map users. Even if users are different from nonusers, there are hundreds of millions of them spread across many regions of many countries. Whether the distinctiveness of our sample is a problem depends on why we are working with the sample in the first place. This leads to the question of purpose.

PURPOSE: WHAT DO I WANT TO LEARN FROM THE SAMPLE?

A deluge of statistics can flow from one sample. Which statistics are relevant and which are distracting depends on your purpose. It is a bad habit to wander aimlessly through sample data. The aimless wanderer can stumble upon important observations, but a walk through the data will be more fruitful if preceded by a reflection on purpose. Before crunching numbers, ask, "What do I think I want to learn from the sample and why?" Having an answer, even if tentative, will focus your efforts and help you gauge the strengths and limitations of the sample.

Your big-picture purpose could be anything—reducing addictions, increasing city revenues, saving the whales—but what you hope to learn from the sample more narrowly probably falls into one or more of the following categories:

- Counts: How much money have investors put in funds formed to take advantage of a federal tax incentive? This would indicate how powerful the incentive is and how much it costs the federal government in forgone revenue.
- Percentages: What percentage of households have student loan debt? This tells us how much of the population would be affected if student loans were forgiven. Percentages can likewise indicate concentration—for example, the percentage of outstanding student debt held by households with an income greater than $250,000 per year.

KNOW YOUR SAMPLE AND DATA   25

- Central tendency: How much does the typical household spend on electricity bills in a year? This is relevant if we want to estimate how much a 10 percent increase in the price of electricity will increase the electricity bill of the typical household.
- Relationships: How does the probability of death from infection vary with age? This would inform decisions about who should receive priority for vaccinations.

As the categories illustrate, we can calculate numerous statistics depending on our purpose. Whether the statistics serve the purpose can depend on their generalizability.

GENERALIZABILITY: DO SAMPLE STATISTICS APPLY TO OTHER GROUPS?

The more generalizable a sample statistic, the more it tells us about people outside of our sample. Generalizability may matter greatly or not at all. A nonprofit organization may primarily care about the behavior of its clients and an elected official about the interests of his constituents. Other audiences may be interested in the entire market or all residents of the state. Even here, a small sample can contain information relevant to broader populations. Data from homeowner-initiated water tests might reveal lead in the water of a large share of tested homes. Even if tested homes are older than other homes, the presence of lead in even a handful of homes would (or should) grab the attention of the city water authority.

The generalizability of a sample statistic depends on two things—the sample and the statistic. Consider the sample from the *Diamond Princess* mentioned previously. Suppose we are in February 2020 and want to learn about the general characteristics of COVID-19, so generalizability is important. The sample of passengers is not representative of the global population in demographic terms, but neither is it a sample of pre-school-aged children or of chain-smoking octogenarians. The passengers are generally older, but age and health status vary among them. Thus, sample statistics might generalize enough to inform policy that covers people who do not go on cruises, especially because so little was known at the time.

The generalizability of a statistic from the *Diamond Princess* also depends on the statistic in question, which is to say, what we want to learn from the sample. If we came from Mars and wanted to know the average age of earthlings, the average age of people on the *Diamond Princess* would greatly overstate the population's average age. But the situation was otherwise in February 2020. Elected officials and public health experts wanted to learn how quickly the virus spreads, the percentage of infections that were fatal, and so on. Although the passengers of the *Diamond Princess* sample were generally older, it is not clear that the virus would spread much faster among older people than among younger people. Moreover, one might explore such a hypothesis with the *Diamond Princess* sample, comparing older and younger passengers. In conclusion, we might learn a lot about the virus by seeing how it worked among this sample and among particular groups within it.

One could raise many objections to using the *Diamond Princess* sample to learn anything generalizable. People do not live on cruise ships, so the infection rate among passengers might be wildly different from in more common conditions. Perhaps. But how different are interactions on a cruise ship from those in a small town where everyone shops at the same grocery store and eats at the same handful of restaurants? Or how different are they from an apartment building where everyone uses the same four elevators and passes through the same common space?

Of course the *Diamond Princess* sample is imperfect, but perfect can be an unhelpful standard. A more useful standard is whether a sample statistic can expand or refine current knowledge. If the *Diamond Princess* had happened five years after the emergence of the virus, no one would have bothered with the ship's data. But at the dawn of the pandemic, it was the difference between flying blind and flying with foggy glasses. Thus, the most relevant standard for the policy aide will often be, Does this improve upon current knowledge? Using that standard instead of a higher one will lead to better information entering the discussion sooner.

Some statistics generalize better than others even though they are calculated from the same sample. The US Department of Agriculture conducts an annual survey to collect information on the characteristics of farms and the people who own and operate them. The survey is long and includes sensitive financial questions. About one-third of sampled farms do not respond, and those tend to be large farms. Using information from another source, my colleague and I found that farms that do not respond to the survey harvest 56 percent more acres and have 74 percent higher sales than respondent farms.[10] Thus, if we want to report on the size of the average farm, the average value from the respondent sample understates the population value if not corrected for the bias introduced by different response rates across farm types.

But suppose we are interested in the relationship between two variables, not in one variable's average value. Would the relationship between a farm's market value and its mix of commodities be fundamentally different between a random sample (all sampled farms) and a respondent-only sample? Not necessarily, and perhaps not often. Our analysis of this and another relationship showed similar relationships across the two samples.[11] In statistical terms, samples can easily give biased estimates of unconditional population values (the average farm size) but reasonably generalizable estimates of conditional values (the average profit among large farms) or relationships between variables (a farm's market value and its commodity mix).

A Summary Example

To tie together the points about sample data, suppose you are a policy aide in the city's Office of Health and the Environment. The city water authority recently had to repair the water lines in one neighborhood and also recently began a practice of testing water quality before and after it does work in an area. So before starting the repairs, it tested the water in all one hundred homes in the neighborhood. The tests showed that five of the one hundred homes had lead concentrations above the action level set by the Environmental Protection Agency.

Is this a random sample? No. The water authority tested homes only in the neighborhood where it planned repairs. The neighborhood might be older than most neighborhoods in the city, its residents lower income, and so on. Do you see how "Is it a random sample?" leads to a dead end? Depending on who you had for a statistics professor, you might abandon the data, throwing up your hands and declaring that we cannot hope to learn anything scientifically useful about water quality in the city from such a sample.

But let's ask the questions about origin, purpose, and generalizability.

The data originated from work on homes linked to a particular water line that needed repair. Because the neighborhood is older, the homes are generally older than the average city home. At the same time, all homes in the neighborhood were tested. Good to know. Now consider your purpose in the Office of Health and the Environment, which is to stay abreast of emerging data and alert the relevant city leadership of issues that warrant their attention. That five homes had high lead concentrations warrants someone's attention. At the very least, the water authority should correct the problem in the five homes. And even though sample homes are older on average, the 5 percent incidence of high lead concentrations suggests that the water authority should test other neighborhoods, perhaps prioritizing older neighborhoods that are likely served by lead pipes.

In summary, the data originated from routine work by the city water authority and reflect a particular neighborhood. Given your purpose, the distinctiveness of the sample does not make it irrelevant to you or the people you support. And although sample values might not generalize to all other city neighborhoods, they raise the prospects of problems elsewhere and the need to expand testing.

Know Your Data

To confidently present statistics to someone who might use them in a decision, know your data. To not mislead others or embarrass yourself, know your data. Be prepared for pointed questions about units or the change in sample size when looking at different variables. And if someone questions the quality or appropriateness of your data, it is not enough to say that such and such person or organization also uses the same data. You must know the data for yourself and defend them on their merits.

Know Key Features

To know your data well, know its key features, including the following:

· *Definitions.* What does it mean to have poor mental health or be food insecure or live in poverty? Interpreting statistics—the poverty rate, for example—requires an understanding of the definitions behind them.

Statistics on poverty and income inequality in particular highlight the importance of definitions, with seemingly small changes in definitions yielding significantly different statistical values. Much has been written on the appropriate definition of income for calculating the poverty rate or measures of income inequality. The Census Bureau's official poverty measure excludes from income the value of health insurance received from the government and other benefits like food assistance.[12] Likewise, the Census Bureau excludes such noncash benefits from income when calculating its main measure of income inequality. It also uses a pretax measure of income.[13] Unsurprisingly, an expanded definition of income and accounting for taxes paid shows considerably less income inequality between the highest- and lowest-income households.[14]

- *Units*. Are things measured in dollars, pesos, inches, meters, acres, or picocuries per liter (a measure of radon gas)? It is easy to forget the name of someone you just met. It is even easier to dive into an analysis and begin sharing findings without firmly grasping what a one-unit increase in the variable means. And knowing units involves more than knowing that the variable is in pesos per household; it involves grasping the real-world meaning of a certain quantity of the variable. Does a Mexican peso buy ten Corona beers or a tenth of a beer? Will 1 picocurie per liter of radiation lead to cancer, or does it take exposure to at least 10 picocuries?

- *Quality*. Conduct smell tests on key variables. Understanding units and their real-world significance will help you spot values that are unrealistic or inconsistent with other data sources. You don't want to present to an audience that includes someone who knows your topic intimately and have her uncover glaring impossibilities or make disconcerting observations for which you have no answer.

Suppose you are studying household electricity consumption using self-reported data on consumption and spending. Dividing spending (dollars) by usage (kilowatt-hours) gives an average price (dollars per kilowatt-hour). How does this average price compare with others sources of retail electricity prices? If the price implied by your data is twice as high as what the local utility company reports as its average price, further digging is needed. Otherwise, one pointed question from the utility company analyst in your audience could cause everyone in the room to doubt every word from your mouth or every statistic from your data. Seek to understand the data well enough to present with confidence before an audience of utility company analysts and billing personnel who deal with data like yours all day long, Monday through Friday.

- *Distributions.* Know more than the mean of your main variable. Just as it would be embarrassing to not know your units or to report implausible values, it would be embarrassing to report that the average house has 5 micrograms of lead per liter in the drinking water—well below the action level of 15 micrograms set by the Environmental Protection Agency—and not know that 80 percent of sampled homes have zero lead and 20 percent have 25 micrograms of lead per liter. Later, we will discuss descriptive statistics in detail, but for now, grasp the need to understand the distribution of values such as the percentage of observations with values above or below some key level, which may be 0 or 15 (the Environmental Protection Agency's lead action level) or something else.

- *Timing.* The measurements in your data correspond to a particular moment (December 31, 2020) or period (January 1, 2020–December 31, 2020), which you should know. Imagine having household wealth data from January 2020, when the S&P 500 stock market index was hitting records, versus having it from March 2020, after it had lost about a quarter of its value. Knowing timing can help you understand why your values might be lower or higher than what the casual observer might expect.

- *Missingness and its treatment.* Missingness is when some people (or farms or stream locations and so on) do not have a value for a particular variable. Entire books are dedicated to the ills and cures of missingness, but the basic issue and potential implications are simple. If a key variable has missing values, you must decide if such observations should be ignored or filled with values calculated from observations with data, a practice known as imputation. In either case, know the extent and pattern of missing data. This involves knowing the percentage of observations that have missing or imputed values: 5 percent may merit only a footnote; 50 percent could doom the endeavor.

Note that what appears to be missing data may not be missing at all. A crop farmer who does not answer a question about antibiotic use in cows should not be treated as missing data—the question is relevant to only some sampled farms. Similarly, data are sometimes collected or digitized such that only nonzero values are recorded, in which case missing cells might properly be interpreted as zeros.

Understanding patterns in missingness is also important for gauging how missingness might bias your statistics. For example, how do people with missing values for income compare to people who reported income? If they are older and better educated, then their missing income is probably higher on average than that of people who reported income. Filling in their missing values with the average income of people of the same age (but not necessarily the same education) will likely cause the sample average income to be less than the average income that would have been observed had everyone reported income.

Know When to Give Up

The rote stricture to avoid all nonrandom samples is bad guidance for the policy aide. Yet sometimes the sample and the data might be so problematic that you are better off abandoning them. The examples in the following sections illustrate how the nature of the sample and the quality of data can make them unfit for a particular purpose. In neither case would uncertainty about the sample or data be addressed by calculating and reporting standard errors and confidence intervals (more on these statistics in chapter 3). Such measures of uncertainty are themselves calculated from the data, and so uncertainty about the sample and data bleeds into uncertainty about estimates of uncertainty. The proverb proves true: garbage in, garbage out.

THE SAMPLE WON'T DO

Sometimes the sample won't serve your purpose. It might not have the scale or representativeness to make statements about the population you care about. Inferring population values is not always the goal, but if it is, the link between the sample and the population is crucial.

During the early stages of the COVID-19 pandemic, the White House Council of Economic Advisers sought out localized and timely measures of severe cases of infections. The number of COVID infections indicated by positive test results depended upon testing effort and availability, which varied greatly over time and space. Confirmed infections also did not distinguish mild cases from severe cases. Hospitals knew how many people they were admitting with COVID-like symptoms, but hospital reporting to the federal government was incomplete and often with a long delay. As an alternative, we explored insurance claim data that would allow us to calculate the number of claims in a given county for hospitalization for severe respiratory illness. The data seemed like a timely way to track severe illness and robust to changes in testing effort and availability.

Initially the data appeared to have the timeliness that we sought, with just a several-day lag in reporting. Closer inspection, however, revealed that while some claims filed on April 12, for example, appeared in the data by April 15, many claims did not appear for up to two weeks. By comparing data sets from different dates—the April 1 data set with the April 15 data set—we could see that what had looked like a two-week decline in claims was actually a delay in reporting. Because timeliness was essential for our purpose, we abandoned the claims data for real-time tracking.

We abandoned the data not because they were bad in the sense that many claims were labeled as involving a respiratory illness when in fact they involved another illness. We abandoned them because of the muddled link between our sample and the population. The whole purpose was to track in a timely manner the number of serious COVID-19 infections in the population. Until enough time had passed for all claims to be processed, we could never know the share of total claims that had been processed in the last week. If we did, and if that were constant over time, we could scale up last week's claims by a certain factor. But without such information, we could not use the claims data to provide an estimate of the number of serious COVID-19 infections in the past week. We could say only that a certain number of claims had been processed for respiratory illness but that the number was an unknown share of the total number of respiratory illness claims. Not helpful.

The link between sample values and population values is especially challenging for voluntarily reported data. By definition, such samples reflect only the people or organizations who chose to report data. Do reporting organizations represent 10 percent of the relevant population or 90 percent? Sometimes it is impossible to know, as an example will illustrate.

The Opportunity Zone provision of the 2017 Tax Cut and Jobs Act gave investors a tax break if they invested in selected low-income neighborhoods through Qualified Opportunity Funds. It was unclear at the time how much investor interest the tax break would attract. The professional services organization Novogradac created a public list of Qualified Opportunity Funds; for this, funds voluntarily reported their data to the list, which Novogradac then compiled on its website. In May 2019, 132 funds reported to Novogradac.[15] Did that mean that 132 funds existed in the entire country? Only if every single fund was reporting to Novogradac. Clearly, Novogradac's count of funds was an undercount of the total number of funds. But did it represent 10 percent of funds or 90 percent? There was no way to know.

Later, the US Department of Treasury reported that about 1,500 funds had declared themselves to be Qualified Opportunity Funds for the 2018 tax year.[16] This means that in early 2019, Novogradac's list represented less than 10 percent of all funds in the population. As such, Novogradac's numbers for the count of funds and the capital they had raised grossly understated how much investor interest the tax break had attracted.

THE DATA ARE GARBAGE

Sometimes the sample might fit your purpose well, but the data are bad. On another occasion at the Council of Economic Advisers, I sought to track housing prices by neighborhood across large parts of the United States. I had access to data on millions of housing transactions, which seemed to provide all I needed for my purpose. The data covered the regions of interest and the variables I needed, and they were timely enough for my purpose. But I soon encountered problems. Many transactions lacked data on the attributes of the house, which were important because housing values are typically measured per square foot of building size. It was not clear why the data were missing for some houses and not others. Even with transactions with square-footage data, the accuracy of the data was suspect. In one sale, the lot size associated with a house was listed as half an acre, for example, but when the house was sold again several years later, the lot size had fallen by a half or doubled! Worst of all, decisions I made about which transactions to exclude from the analysis often led to wildly different values for statistics of interest.

The general problem created by poor measurement is that it makes the data and the resulting statistics unreliable and even worse than having no data. Poorly measured data can take us from not knowing where to go to being confident about going in the wrong direction. To illustrate, consider that housing sale prices can change from one quarter to the next because of changes in the types of homes sold in each period. The mean house last quarter might have had 1,500 square feet compared to the 3,000 square feet of the mean home this quarter. Putting prices in terms of dollars per square feet addresses the change in composition, but poor measurement of square footage can make the per-square-foot price unreliable. Suppose a 3,000-square-foot home sold and was wrongly reported to have half as much square footage. The halving of square footage will double the price per square foot, introducing variation in prices driven by mismeasurement. Depending on the extent of the error, it could make it look like housing values in a neighborhood are decreasing when in fact they are increasing, or vice versa. Moreover, error in measurement hides real correlations among variables, making it appear that they do not move together when they really do.

Poor measurement can also affect interpretability. Imagine that your organization sought to understand its clients better and on one of its forms asked, "What is your estimated income?" Think about that question. How would you answer? Would you give your monthly income or your annual income? Your individual income or your household income? Your take-home income (after taxes and deductions) or your gross income? Two people, each with $100,000 in annual household income, could report very different numbers depending on whether they reported individual or household income or annual income or monthly income.

It will not help to inspect the income data more closely because there is no way to identify with confidence who reported individual income and who reported household income. The resulting statistics will therefore be uninterpretable. Imagine reporting to the organization's board that the average client has an income of $35,000 and being asked how that compares with the average income of people in the county. The comparison is meaningless because the county average income is for a year and an entire household, but your $35,000 reflects a tossed salad of different income concepts. Again, having such unreliable data can be worse than having no data.

GUIDANCE FOR KNOWING WHEN TO GIVE UP

Do not give up simply because sample data surprise you. We are incapable of learning if we blindly reject everything inconsistent with our prior understanding. Rather, give up on a sample or its data if they show themselves to be irrelevant or unreliable for your purpose. Regarding the sample, the questions about origins, purpose, and generalizability are good ones. Related questions to ask about the sample include the following:

- Am I interested only in population values? If so, is the link between the sample and the population clear? Recall that some population values are harder to infer than others. Totals and unconditional means are generally more sensitive to sample composition than a correlation coefficient between two variables.
- Have definitions or procedures used to determine the sample changed over time? If so, think through their implications. For example, samples from administrative data can evolve as governments revise regulations about who is required to complete various forms. If the composition of businesses in the sample changes dramatically over an important period for your analysis, consider giving up.
- Is the sample interesting in itself because of its size or its relevance to a particular policy? You might abandon the sample for one purpose but realize that it could be useful for another.
- Could you access other samples that might have better coverage or more details?

Questions to ask of the data include the following:

- Are the definitions of key variables clear and consistent over time? If you want to look at changes over time and a large change in the definition of the variable occurred between periods, consider giving up if the change is substantial.

- For self-reported data, do respondents have the incentive to misreport? Businesses completing a survey about the expected costs of complying with a potential regulation have the incentive to report worst-case scenarios.

- Do implausible values commonly appear, or do the data vary erratically over time in strange ways? One way to learn about the quality of your data is to think of areas where it overlaps with other reliable sources. For example, you might have self-reported electricity bills and can compare the sample average annual bill with that reported by the regional utility company in its annual report. How do the two compare in one year and over time? Perhaps self-reported bills are always higher, but the year-over-year changes are similar.

More generally, a great practice for thinking through the suitability of the data is to talk to people familiar with them and ask for their thoughts on using the data for your purpose. If using Google mobility data, get the Google employee who compiles the data on the phone. People in the weeds of data generation often love to talk about it. When I worked at the Department of Agriculture's Economic Research Service, I loved when analysts called me with questions about data collection and implications for this or that statistic. It was an encouragement to know that people were using the data that we worked hard to provide. Be aware, however, that people closest to the origins of the data are usually the most pessimistic about their quality, just as people who work in restaurant kitchens are often most critical of their own restaurant's food. But stay focused and listen carefully for how their concerns relate to your purpose for the sample and data.

3

Know Simple Statistics and Their Power

Simple statistics are summaries of data that can be explained in one sentence to a broad audience. They are unadorned totals, percentages, means, percentiles, ranges, and simple measures of linear relationships. They may suggest causal relationships—they may show that people eligible for unemployment benefits are less likely to be employed than those ineligible—but their purpose is not to isolate the effect of benefits on employment apart from confounding factors. It is to establish basic facts like the number of people who applied for unemployment benefits last week.

Despite their seemingly modest purpose, simple statistics are often the most relevant and helpful for people making decisions. As discussed in chapter 1, statistics can be decisive for policy by revealing whether a problem is real and merits attention. Consider the chapter's example of a city whose buses were supposedly arriving late to their stops. A simple statistic was enough to keep a media story on tardy buses from absorbing the attention of the city's transportation leadership: the percentage of stops where the bus arrived late showed that there was no problem with the bus system.

An example from the real world further highlights the power of a simple statistic or two. In April 2019 the Department of Commerce submitted a report to the White House recommending that the president require that US nuclear-power producers purchase a large amount of uranium from US mines. According to the report, the domestic purchase requirement would further US national security interests and have little effect on power producers and their customers. The department asserted that at $55 per pound, an arguably modest price, US uranium producers would produce the 6 million pounds required by the recommended policy.[1]

While others in the White House questioned the department's national security argument, the Council of Economic Advisers cast doubt on its economic assertions using simple statistics. In recent history, the annual average price for uranium had exceeded $55 price in multiple consecutive years, and yet US production never came close to 6 million pounds.[2] This suggested that a much higher price would be needed for US producers to bring 6 million pounds to the market. By extension, an already-struggling nuclear-power industry would face higher costs, which it would pass to households and that in some cases might force plants to close. Concerns about the cost increase entered the decision memo to the president, who ultimately rejected the recommendation from Commerce.

The right simple statistic can also permit a credible real-time estimate of the likely effect of a current policy proposal or event. If the president of the United States is considering closing the border with Mexico out of frustration with illegal immigration (which he did[3]), a key statistic for understanding the scale of economic disruption is the value of goods and services that cross the border in a typical day. Or if suspected Iranian attacks on ships in the Strait of Hormuz threaten to close off passage to oil tankers (which they did[4]), what share of the global oil trade would be affected?

Simple statistics also permit a better understanding of our organization, our city, our watershed, our country, or our world. The understanding then colors how we interpret new facts and emerging issues. Knowing the likelihood that poor kids from a particular area will rise above their parents' economic condition may not drive any one policy decision at the moment, but it will be in the mind as people consider proposals for geographically targeted tax breaks or education grants.

Before you skip this chapter because you learned about the mean in fifth grade, be warned that simple is not always straightforward and obvious. For example, a thorough understanding of the definition of simple statistics is key in interpreting them and knowing what they can and cannot say. And just as each of the carpenter's tools has a purpose and occasion, so it is with simple statistics. Each has its particular strengths and limitations. We now turn to understanding them.

Central Tendency: The Typical Household Has 2.9 Cars

The Mean

The simple arithmetic mean is the ratio of two totals. The mean number of cars per household is the total number of cars owned by all sample households divided by the total number of sample households. Aside from the total, it is perhaps the most widely recognized and used statistic and the most common measure of what is typical. Note also that a percentage, such as the percentage of households with a car, is a mean. It is the ratio of two totals: the number of households meeting a condition, such as having a car, divided by the total number of households.

Seeing the mean as a ratio of two totals makes it easier to see its strength: every household participates in influencing the mean. Every household increases the denominator (the total number of households), and most affect both the numerator and the denominator. Changing one household's value will therefore change the mean. In this sense, the mean seems inclusive and democratic, with everyone getting a vote.

Yet one household might have a million more votes than another household. Households with large values can influence the mean greatly. Imagine studying car ownership and having a sample of ten households, nine of which have one car each and one of which collects and fixes up cars and has twenty of them. The mean number of cars per household is 2.9 cars, but if the outlier household had one car like the rest, the mean would be 1. If someone proposed a per-car tax, the average tax owed (based on the average number of cars owned) would be nearly three times larger than what 90 percent of the households would actually pay.

Or consider the town of Medina, Washington, with 1,195 households that have a mean income of $323,190, according to the US Census Bureau's 2015–2019 estimate.[5] As a group, they have $386 million in annual income (1,195 × $323,190). The total apparently does not include one of the town's most notable residents, Jeff Bezos, the founder of Amazon. Evidence suggests that he makes about $1 billion in taxable income per year, in which case his inclusion in Medina's average income would increase it from a very comfortable $323,190 per household to a very, very comfortable $1.2 million.[6]

Another insight regarding the mean is that it can be calculated from different perspectives. Consider how much state governments tax oil production in the United States. We can provide at least two mathematically and conceptually distinct perspectives on the mean— the perspective of the state and the perspective of the barrel of oil.

For the state perspective, take the tax rate from each state and average across states, treating each state alike regardless of how much oil it produces: the rate of Texas receives the same weight as Indiana's (a weight of 1). This mean rate, 3.6 percent, according to one study, tells us that the average state receives $3.60 in revenue for every $100 in oil produced from within its borders.[7]

The 3.6 percent rate describes what the typical state does. To know what tax rate is applied to the typical barrel, we must treat Texas and Indiana differently because many more barrels face the Texas rate than the Indiana rate. For the rate from the barrel's perspective, sum oil tax revenues across all states and divide by the sum of oil production across all states. This approach gives Texas more say in influencing the mean rate. The calculation is in fact a weighted average rate, where each state's weight is its share of total production across all states.

The same study that estimated that the average state taxes oil at 3.6 percent found that the average barrel is taxed at 4.2 percent, which is 17 percent higher than the average state tax rate. If we want to describe the state tax revenue generated by the typical US barrel, we need the perspective of the barrel. If the United States produces $200 billion worth of oil and we assume the average state rate, we wrongly conclude that oil generates $7.2 billion for state governments (= 3.6 percent × $200 billion) when in fact it generates $8.4 billion (= 4.2 percent × $200 billion). This is because more oil production occurs in states with above-average tax rates.

The Median

The median is the next most commonly used measure of central tendency. Roughly speaking, it is the value such that 50 percent of observations have a lower value than it and 50 percent have a higher value. Unlike the mean, the median is unaffected by observations with extreme values, positive or negative. If Jeff Bezos's income goes from $1 billion to $1 million, it will not affect median income in Medina, which was about $212,000 per household for the 2015-2019 period.[8] By comparison, it is possible that 90 percent of Medina households have less than $1.2 million, the city's mean income if the Bezos household is included in the mean.

The median's robustness to outliers is nice but comes at the cost of insensitivity to changes in parts of the distribution that we might care about. Much change can happen at either the lower or upper end of the distribution without changing the median. Recall the car example in which nine households had one car and one household had twenty cars. The median number of cars per household is one car. This is a good reflection of central tendency in this case because most households have exactly one car. But imagine that four households were robbed of their car. The loss does not affect the median, which is still one, even though 40 percent of households now have zero cars. Moreover, we could redistribute the four robbed cars to three other households in the sample, and the median would still not change!

The Mean Matters for Budgets and the Median for Fairness

The different mathematical properties of the mean and the median map onto different policy priorities and uses. Broadly speaking, the mean is relevant when the interest is in aggregate outcomes, and the median is relevant when the interest is in knowing what is happening to the person or place in the middle.

If our concern is the growth of the overall economy, the overall catch in the fishery, or the balancing of the city's budget, the mean is of primary interest. In such cases, the distribution of values across places, boats, or infrastructure projects is secondary. The city budget manager cares little that every infrastructure project is right on budget or that some are over budget and some under budget. His primary concern is that the average project is on budget. If it is, total spending across all projects will equal the total amount budgeted because of how the mean is defined.

The budget example illustrates how the mean's sensitivity to outliers can be a feature instead of a bug. When one project has a massive cost overrun, it does not matter if most projects are a little under budget; spending on the outlier will blow the budget. One can imagine similar priorities in other settings. Consider managing the Alaskan crab population, for which biologists have determined the number of crabs that can be caught in a season and still leave enough to stabilize the population from year to year. From an ecological perspective, it does not matter if one boat catches all the crabs. The key is that the average amount caught per boat matches the average quota, which is to say that the total catch matches the total allowable catch.

In other settings, we are concerned about the experience of those in the middle. Is policy or a changing economic wind treating them fairly? The budget manager might actually care that the median project and contractor perform under budget and that one outlier contractor is offsetting the good work of the median contractor. Similarly, the crab manager might have a mandate to pursue multiple goals, including supporting a fleet of locally owned and operated crab boats. The experience of the median boat would therefore matter, as it could reflect a shift toward only a few large corporate companies catching all the crabs.

Whether interested in aggregate outcomes or individual experiences, it is important to understand how values are dispersed across people, places, or projects. As discussed in chapter 2, understanding the distribution of key variables is part of understanding your data. It aids in smell checking the data and understanding the magnitude of a one-unit change in the variable.

Dispersion: No One Has 2.9 Cars

The mean and the median values are useful, and people like having one number to present as "typical" or "average." The problem, which the policy aide should always remember, is that average is often not average: the arithmetic mean is often not typical. With the sample of ten car-owning households, the mean household has 2.9 cars even though no household has 2.9 cars or even two cars or three cars. In fact, 2.9 cars is nearly three times more cars than what 90 percent of households have and dramatically less than what 10 percent of households have.

The median may not help either. In 2020, farm households had a median farm income of negative $1,200, meaning that the median household lost money from its farming activities.[9] The number begs further questions: How many households turned a profit from farming? How many earned more than a few thousand dollars? Anyone seeking understanding, not just a statistic, must explore the dispersion in the data.

Ranges and Percentiles

A natural first step to exploring dispersion is to look at the highest and lowest values for key variables, which is the range. Looking at the range is a data-quality check; if the oldest student in your sample is age 170, there was a problem with data reporting or recording. For unfamiliar units—picocuries per liter—the range furthers our understanding of the diversity of experiences in the sample. It is, however, only a small step toward understanding dispersion.

Looking at percentiles and the range between them reveals more about dispersion. The 50th percentile is the value such that half of all households have a value equal to or less than the value, so it is the median value. The values at the 25th and 75th percentiles bound the middle 50 percent of sample observations. That is, 50 percent of households have a value above the 25th percentile and equal to or below the 75th percentile. This range is known as the interquartile range.

Another helpful range is from the 5th percentile to the 95th percentile. It spans the middle 90 percent of households and is useful for discussing what is *not* typical, which can be less prone to misunderstanding than measures of what is typical. A household with a value outside this range is among the top 5 percent of households or among the bottom 5 percent. Note that nothing prevents the mean from falling below the 5th percentile or above the 95th percentile. A massive loss from one very large farm could push the mean farm profit below the 5th percentile, whereas a massive profit could push it above the 95th percentile.

Percentages

Understanding dispersion is more than simply reporting ranges or percentiles. It is about understanding important thresholds in the data and how households (or places or farms) are distributed around the threshold. Percentages are helpful here. They tell us which proportion of households meet a certain condition, for example, the percentage of households in the interquartile range, which is always 50 percent by definition. But aside from marking a place in the distribution of households, there is nothing particularly meaningful about being in the interquartile range or in the top 25 percent. Other values can have inherent significance.

Zero is often an important value, separating the haves from the have-nots. Owning or not owning a car, for example, means depending on public transportation, friends, or ride-hailing services to get groceries. In epidemiology, the mere presence of a virus in a school is consequential. The same is true in ecology and the presence of invasive species like the Asian carp. Particular nonzero values also carry real-world weight. Particular incomes mark the difference between eligibility and ineligibility for health-care subsidies or housing vouchers. The Environmental Protection Agency has set thresholds to separate safe and unsafe levels of lead in drinking water and radon inside homes. Hog farms above a certain size are subject to more stringent environmental regulations. You get the idea.

We often also care about concentration, which can be measured with a percentage like the percentage of government-provided student loans held by wealthy households. Percentages showing concentration matter for policy because they indicate how much of the benefits or costs of a policy decision will accrue to a particular group. As an illustration, one report found that the wealthiest 20 percent of households held nearly one-third of federal student loan debt while the least wealthy 20 percent had only 8 percent.[10] Blanket forgiveness of debt would therefore disproportionately benefit wealthier households.

Note that a common error for the beginner and the advanced is mislabeling the difference between two percentages, say, 5 percent and 10 percent. It is a difference of 5 percentage points, not a difference of 5 percent. That is, 10 percent is actually 100 percent greater than 5 percent.

Variance and Standard Deviation

The variance and the standard deviation also measure dispersion, although they are not nearly as simple or easy to use in a sentence as ranges and percentiles. The sample variance measures how values are dispersed around the sample mean value. Explaining the five-step calculation of variance builds intuition for its interpretation. First, calculate the mean. Second, find the difference between the mean and each household's actual value. Third, square each household's difference. Fourth, total the squared differences across all households. Last, divide the total by the number of households. Summarized in mathematical notation, the variance of variable x in a sample of n households is

$$\text{Variance of } x = \frac{\sum_{i=1}^{n}(x_i - \bar{x})^2}{n - 1},$$

where i refers to a particular household. Having more households far from the mean (above or below it) increases the variance by increasing the difference with the mean, the squared difference, and then the total of squared differences. Thus, less dispersion means a smaller variance, which can be as low as 0 if all households have the same value.

But the variance is hard to use in a sentence. It is not in the original units because of squaring the differences. A step toward greater interpretability is to take its square root, which gives the sample standard deviation. Now we are back to the original units, say, cars owned per household. The standard deviation tells how much a household's number of cars will differ, on average, from the mean number of cars. This can be confusing because the mean household has the mean number of cars, so it seems like the mean difference should be 0. In fact, this is true: the mean difference from the mean is 0 because differences below the mean exactly cancel out differences above the mean. But the canceling out does not happen when the difference is squared because squaring makes negative differences positive.

To understand the standard deviation, imagine blindly grabbing a particular household from the data and guessing that it has the sample mean number of cars. In many instances, our guess will differ from the household's actual number of cars. The standard deviation measures how bad the strategy of guessing the mean is on average. For the sample of ten households, with nine households with one car each and one household with twenty cars, the standard deviation is 6.0 cars. So, if we grab a household at random and guess that it has the mean number of cars (2.9 cars) and do this over and over, we might expect our guesses to be off by six cars on average.

As you might intuit, the standard deviation, being based on the mean, has the same strength and weakness as the mean. The strength is that every household participates in voting for the standard deviation, even if it is far out in the tails of the distribution. This is the case for the household with twenty cars. In a sense, this is good because it alerts us to considerable dispersion of some sort. If we in fact draw a household at random and guess that it has the mean number of cars, we will be off by a lot on average (six cars off!). This is good to know. Yet the strength is also the weakness. In the car sample, the standard deviation suggests considerable dispersion in car ownership across households when in reality there is dispersion only in the sense that one household had dramatically more cars than any other household. If we select that household and use the average as a prediction, we will be off by a whopping seventeen cars. But setting the outlier household aside, there is no dispersion: everyone has the same number of cars! And if the outlier had two cars instead of twenty, the standard deviation would be 0.3 cars instead of 6.0 cars.

Sometimes people transform variables to put them in terms of standard deviations, usually by taking each household's value, subtracting the mean value, and dividing by the sample standard deviation. So, a household whose transformed value is 0 has the mean value, and the household with a transformed value of 1 is one standard deviation above the mean. The goal of the transformation is to improve comparability of variables with different units. So we can talk about a one-standard-deviation difference in a student's class size, in a student's parent's income, or in a student's shoe size.

The standard deviation provides some help in gauging magnitude and dispersion, but it is a disaster for communicating with most audiences. I may go as far as saying that the standard deviation should not cross your lips outside of wonk circles. Most of the public and most people in leadership at the nonprofit or the state government will not be helped by it. It will be distracting and even off-putting in the same way that it would be off-putting if the biologist being interviewed by the local news reporter said that numerous *Odocoileus virginianus* are damaging the ecosystem of the state's forests.

Only the most statistically minded think in terms of standard deviations. People get it when you say that 90 percent of men are between five and six feet tall; they don't get that 90 percent of men are within

1.8 standard deviations from the mean. Although I teach statistics, I still have to translate in my head when I speak of standard deviations. It is like speaking in a foreign language. Imagine the city council member who is told by the analyst from the Office of Health and the Environment that air pollution in the city this summer has been 0.5 standard deviations about its historical average. Come again?

Yet some statistical concepts are important even if they are not readily (or ever) grasped by more general audiences. The standard deviation in particular is central to inferring certain population values from sample values. Sometimes our purpose is such that we are interested only in the households in our sample. Other times we are interested in the sample only so far as it can tell us about a larger group. It is to this case that we now turn.

Sampling Error and Population Values: Maybe It's 5.9 Cars?

Sometimes our data contain all the places or people of interest. We may have data for all our customers, clients, or property-tax payers. But other times our sample includes only a sliver of the population of interest. Our sample of ten car-owning households may have value only if we can use it to infer the total number of cars owned by county residents and therefore estimate how much revenue a car tax might generate. If so, the "Is my sample random?" question becomes important. Suppose the ten households come from a poorer part of the county. If having more income is associated with owning more cars, our sample mean might greatly underestimate the county population mean and, by extension, the revenue that a car tax would raise.

The example highlights the value of having a random sample. By definition, we would expect that households randomly selected from the population would on average have characteristics similar to those not selected—similar incomes, demographics, and number of cars. This means that our method should on average give the true number of cars per household in the county as a whole. We therefore have a basis for multiplying the sample mean by the total number of households in the county to estimate the total number of cars in the county. We could then estimate future tax revenue by multiplying the estimated total number of cars by the proposed per-car tax (assuming that the tax doesn't cause people to sell their cars).

But before sharing the revenue estimate with the county budget director, understand what it means that "our method should on average give the true number of cars per household in the county as a whole." It means that if we keep drawing random samples and calculating the mean for each one of them, the mean of the sample means should match the population mean. For any given sample, however, selected households might look different from the rest of the population. Recall that our sample of ten households has one household with twenty cars, which inflated the mean. The real frequency of such a household in the population might be one in one hundred, not one in ten, as suggested by our sample. If we drew another sample of ten households, in place of the twenty-car household, we might get a two-car household, which would yield a dramatically lower estimate of the population mean number of cars per household.

The Standard Error

The possibility of getting a very different answer from a new sample should make you uneasy about spreading your revenue estimate around the county government and local media. It would be nice to quantify how much the car estimate, and therefore the revenue estimate, might differ from the population value due to sampling error. This is what the standard error of the mean quantifies. Just like the standard deviation measures how much individual household values differ from the mean on average, the standard error measures how far individual sample means differ from the population mean on average. This allows us to estimate a confidence interval for the population mean. The confidence interval gives a range that likely includes the population mean. It reflects the reality that we have only a sample, not the entire population, and that the sample itself might not accurately reflect the population.

Standard errors and confidence intervals are hard to explain in a sentence and so do not fit the definition of a simple statistic. Moreover, some decision makers will care nothing for them, especially if they are just looking for a talking point. It is nonetheless your job to know the uncertainty around your simple statistics that are estimates of population values.

Before we can calculate a confidence interval, we have to calculate the standard error. The estimated standard error of the mean is simply the estimated population standard deviation divided by the square root of the sample size:[11]

$$\text{Est. standard error of the mean} = \frac{\text{Est. population standard deviation}}{\sqrt{n}}.$$

Or alternatively, it is the square root of the estimated population variance divided by the sample size:

$$\sqrt{\frac{\text{Est. population variance}}{n}}.$$

It is worth reflecting upon the standard error because it is hard to understand. And if you can explain it to the average person in a sentence, you will appear to have (and arguably have) a statistical savvy in the 99th percentile. The standard error seeks to describe the distribution of many (hypothetical) samples. This distribution, known as the sampling distribution of the mean, is what would result if we drew many samples of a particular size and plotted the distribution of their means on a graph. The standard error measures the distribution of these sample means around the true population mean. In this sense, it is a standard deviation not of individual household values but of sample mean values.

Anything that reduces the standard error makes us more confident that the sample mean will be close to the population mean. A larger sample size reduces the standard error because it increases the denominator. Intuitively, the more households in our sample, the fewer surprises we should expect from another sample of the same size. We have sufficient chances to draw outlier households in general but also to draw them with roughly the same frequency with which they appear in the population. In the extreme, as the sample size grows larger and larger, the standard error approaches 0 because the sample becomes the population.

A larger standard deviation, in contrast, gives a larger standard error because it is in the numerator. Intuitively, more dispersion across individual households (a higher standard deviation) brings more uncertainty over whether our sample is reflective of the broader population. The sample of car-owning households has a large standard deviation, a reflection of the twenty-car household inflating the mean and therefore the deviations around the sample mean. Having drawn one sample with an outlier, we might expect the next sample's mean to differ considerably from the first sample's mean. If the second sample does not draw a many-car household, the mean will be much lower. In contrast, the second sample might include several of the many-car households, which would yield a much higher mean. In short, greater dispersion in the sample implies greater dispersion in means from sample to sample, which means that any given sample mean might be quite far from the population mean.

The example also illustrates another important point: the sample estimate of the population standard deviation is just an estimate. If the original sample excluded the twenty-car household—and such households are a real feature of the population—we would greatly underestimate dispersion in the population and be overconfident that the next sample would yield a similar mean.

The Confidence Interval around the Mean

The sample mean, which is an estimate of the population mean, and the standard error are most of what we need to estimate a confidence interval for the population mean. To construct the interval, we need an upper and lower value in the sampling distribution of the mean. For a 90 percent confidence interval, we need the upper and lower value such that 5 percent of sample means exceed the upper value and 5 percent fall below the lower value. (For a 99 percent confidence interval, replace 5 percent with 0.5 percent in that sentence.) According to the central limit theorem, as the sample size grows, the sampling distribution of the mean becomes approximately normally distributed regardless of whether the raw data (cars per household) has a normal distribution. This means that with large enough samples, 5 percent of sample means are 1.65 standard deviations above the population mean and 5 percent are 1.65 standard deviations below it. Recall that the standard error is simply the standard deviation of the sampling distribution of the mean.

For our car example, the estimated population standard deviation is 6.0. The standard error of the mean is therefore 1.9 (= $6.0/\sqrt{10}$), and the 90 percent confidence interval is as follows:

90% confidence interval = Mean ± standard error · critical value

= 2.9 ± 1.90 · 1.65

= −0.2 cars to 6.0 cars.

This confidence interval fosters little confidence. It roughly includes plus or minus twice the mean of 2.9 cars, a wide range. The wide range, in turn, gives a wide range of revenue estimates, with the estimate equal to the product of the estimated number of cars per household, the number of households, and the proposed county tax per car. Taken literally, the lower bound of the interval suggests a loss in revenue from the tax (a reflection that the normal distribution, which the critical values are based on, is not truncated at any value). Given the range, you might want to pursue a larger sample before talking to the county budget executive.

When we seek to use a sample to estimate a population mean, having a sample randomly selected from the population of interest is key. If the sample is drawn from just one social network or neighborhood, the sample mean is likely far off the population mean, and we won't be helped by calculating a confidence interval. Even if the sample standard deviation is a spot-on estimate of the population standard deviation, the confidence interval will be wildly off because it is centered on the far-off sample mean.

This approach to confidence intervals is known as the frequentist approach. It takes the population mean as a fixed number and all statistics calculated from the sample, including the confidence interval, as random variables. The approach is attractive because the sample data give us everything needed to infer the population value. The downside is that its confidence intervals are hard to interpret and communicate. Think of the confidence intervals as randomly bouncing around from sample to sample while the true population mean remains fixed. Any given confidence interval might be way off and not actually contain the population mean. It is not correct, then, to say that the confidence interval you just calculated contains the population mean with a 90 percent probability. The population mean is a fixed number, and any given confidence interval either contains it or excludes it. What a 90 percent confidence interval really means is that the method to generate confidence intervals generates them such that on average nine of every ten intervals contain the population mean. This is hard to grasp, and most people will think of your 90 percent confidence interval as a range that contains the population value with a 90 percent probability.

The Bayesian approach, in contrast, treats the population mean as a random variable and assigns probabilities to its taking on a particular range of values. This is easier to interpret and is what many people think a frequentist confidence interval provides. The easier interpretation comes at the cost of needing to initially assume the distribution that the random population value comes from, what is known as the prior probability distribution—or simply, the prior.

Relationships

Simple statistics include measures of relationships between variables. Apart from claims about causality, it can be helpful to know which variables move together either in the same direction or in opposite directions. Are older houses more likely to have lead in their water? Even if age is not causally related to lead concentrations, the relationship may still aid in directing resources to houses or neighborhoods more likely to have a lead problem. More broadly, the predictive power of a variable can be useful even if the thing doing the predicting (house age) does not actually cause the thing being predicted (probability of lead). Lead pipes may simply be more common in older homes.

Another example of using statistical relationships is for identifying patterns in who benefits from a particular program. The Economic Development Administration of the US Commerce Department might want to know the poverty rate of areas that have and have not received any of its economic development grants. The point is not that the poverty rate causally affects grant allocations (although it may) but that the administration wants to know if its process for awarding grants incidentally favors certain areas.

Group Comparisons

Comparing measures of central tendency—usually the mean—across groups is arguably the most common approach to describing relationships between variables. Creative and thoughtful group comparisons, often shown in figures, are an essential tool for the policy aide as she seeks to communicate statistical relationships to broad audiences.

Consider measuring economic mobility across generations by relating a person's annual household income as an adult, say, at age forty, with his or her parents' household income when the parents were forty. To see how parental income relates to child income, split people into two groups, those with above- and below-median parental income. Next, calculate and compare the mean child income across the two groups. We might expect children from higher-income parents to have higher average income as adults, but is it 10 percent higher or 100 percent higher than that of children with lower-income parents?

More can be learned by dividing people into finer groups. We could divide them into deciles based on parental income—is their parental income among the lowest 10 percent, the lowest 10 percent-20 percent, and so on? The mean of child income across the ten groups reveals whether people in higher deciles of parental income generally have a higher mean child income, but it can do more. It can reveal kinks and curves in the relationship. We might see that people in the first decile of parental income have a mean child income similar to those in the fifth decile, but among those in the top several deciles, we might see that higher parental income is clearly associated with higher child income. This would suggest that parental income is not a barrier to moving into the middle class, but it is a barrier to moving to the top of the income distribution. As a bar chart, such a figure would transparently reveal a rich story accessible to a broad audience.

The Correlation Coefficient

The correlation coefficient is another common measure of the relationship between two variables. Continuing with the example of parental income and child income, the correlation coefficient between the two is a measure of how much the two incomes vary together (covariance), divided by a measure of how much they vary in general. Mathematically, it is the covariance of parental income and child income divided by the square root of the product of the variances of each:[12]

$$\frac{\text{Covariance }(PI, CI)}{\sqrt{\text{Variance }(PI)\cdot\text{Variance }(CI)}} = \frac{\sum_{i=1}^{n}(PI_i - \overline{PI})\cdot(CI_i - \overline{CI})}{\sqrt{\sum_{i=1}^{n}(PI_i - \overline{PI})^2 \cdot \sum_{i=1}^{n}(CI_i - \overline{CI})^2}}.$$

To be clear on notation, the i subscript refers to a particular child, so child i's income as an adult is CI_i. There are n people in the sample.

The correlation coefficient captures whether two variables move in the same direction (a positive correlation) or in opposite directions (a negative correlation). The denominator is the product of two positive numbers (variance is always positive), so the numerator gives the correlation coefficient its sign. Suppose that the two move in the same direction. When parental income is above average, child income also tends to be above average, and the numerator then sums the product of two positive numbers, and the whole expression is positive. Similarly, when parental income is below average, child income tends to be below average, and the numerator sums the product of two negative numbers, which yields a positive number. A negative correlation coefficient arises when above-average parental income is associated with below-average child income, in which case the product of the two gives a negative number.

The correlation coefficient also says something about the intensity of the relationship. Think of the variation of the child's household income as having two parts, the part that varies with parental income and the part that does not. The part that varies with parental income can never be greater than the sum of the two parts, the total variance, which is measured by the denominator. This is why the correlation coefficient falls between 1 and −1. The fact that the denominator measures total variation also gives the correlation coefficient intelligibility. It presents the shared variation of two variables as a fraction of their total variation, with a 0 correlation coefficient indicating no co-movement and a −1 or 1 correlation coefficient indicating perfect co-movement.

Although a correlation coefficient near 0 suggests no relationship between two variables, it does not capture nonlinear relationships well. Two variables could have a 0 correlation coefficient even though they have a strong nonlinear relationship. The relationship between parental and child income could theoretically be an upside-down U shape, with higher parental income leading to higher child income over a range of low incomes but higher parental income leading to lower child income over a different range of high parental incomes. The correlation coefficient can be 0 or nearly 0 in such a case despite considerable co-movement between the two variables.

By measuring shared variation as a fraction of total variation, the correlation coefficient measures dispersion around a line. The two panels in figure 3.1 have the same best-fit line (discussed in the next section), defined as Child income = 41 + 0.90 · Parental income. In panel A, however, the data fit less tightly around the line than they do in panel B. The correlation coefficient captures this: the data in panel A have a correlation coefficient of 0.57 compared to 0.97 in panel B.

Just as the same best-fit line can have different correlation coefficients, data with different best-fit lines can have the same correlation coefficient. In figure 3.2, panel A, the best-fit line implies that each $1 more in parental income is associated with $0.90 more in child income. In panel B, it is associated with $0.09 more in child income. In other words, the best-fit line in the first panel is ten times steeper than the line in the second panel. The dispersion of data around the best-fit line relative to the total dispersion, however, is the same in both cases, resulting in the same correlation coefficient. As long as the relative dispersion of data around the line is the same, two lines with very different steepness (slopes) will have the same correlation coefficient.

The examples show that the correlation coefficient measures fit around a line but not how much child income changes on average with a $1 increase in parental income. To distinguish a $0.90 increase in child income from the $0.09 increase, we must look to the slope of the best-fit line itself.

The Slope Coefficient

The names of the correlation coefficient and the slope coefficient suggest that they are very similar, but they are only cousins, not sib-

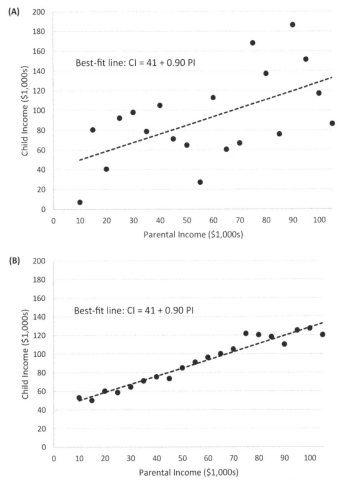

Figure 3.1 Same line, different fit. (A) Less fit: correlation coefficient of 0.57. (B) greater fit: correlation coefficient of 0.97.

lings. What puts them in different nuclear families is what they communicate about a relationship. The correlation coefficient is unitless, ranges from –1 to 1, and cannot reveal *how much* a person's household income increases for each $1 increase in his parents' household income. The slope coefficient, in contrast, can go to negative infinity or positive infinity. It can say that each $1 increase in parental income is associated with $2 in greater household income for the children.

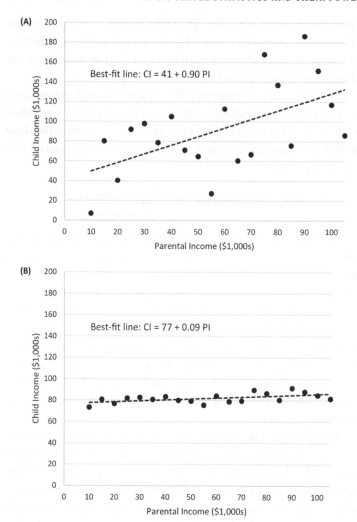

Figure 3.2 Different line, same fit. (A) A steeper best-fit line: correlation coefficient of 0.57. (B) A flatter best-fit line: correlation coefficient of 0.57.

A small change in the correlation coefficient formula turns it into the formula for the slope of a line, also known as the slope coefficient. By "slope coefficient," I mean the β term in the equation of the line relating parental income and child income:

$$\text{Child income}_i = \alpha + \beta \text{ parental income}_i,$$

where β is the slope of the line, giving the change in child income for a $1 change in parental income. (The term α is the value at which the line cross the vertical, or *y*, axis, which is the value of child income when parental income is 0.)

The small change is removing the variance of child income (and the square root symbol) from the denominator of the correlation coefficient formula, which gives the formula for estimating the slope coefficient (where the hat on β indicates that it is an estimate of a population value):

$$\text{Est. slope coefficient} = \hat{\beta} = \frac{\text{Covariance}\,(PI, CI)}{\text{Variance}\,(PI)}.$$

The change has profound implications. The covariation of the two variables was previously divided by their combined total variation, which gave a unitless fraction of shared variation to total variation. Dividing the shared variation by only the variation in parental income gives a statistic with a fundamentally different interpretation. With some calculus that is in most statistics textbooks, it can be shown that this formula gives the slope of a line, which has a clear interpretation: the dollar-value change in child income associated with a $1 change in parental income. Thus, the slope coefficient will be different for the two scenarios shown in figure 3.2 even though both have the same correlation coefficient.

There is more. The formula gives us the slope not of any old line but of a particular line—the least-squares regression line. Among all lines, the least-squares line best fits the data according to the least-squares criterion of minimizing the sum of the squared differences between the actual data points and the line fitted to them. In simple terms, the line cuts the data such that the data points above the line are equally above the line as points below are below it.

As with the correlation coefficient, which has the same numerator, the slope coefficient is based on the mean and has all the strengths and weaknesses of a mean-based statistic. It is influenced by all data points, which is generally a good thing, but it can be very sensitive to outliers, which is often a bad thing, because it means that the relationship might vary greatly from sample to sample.

But the slope coefficient has its own inability to distinguish distinct cases. Two samples can yield the same best-fit line but differ greatly in how well the line fits the data.

Measuring the Fit of a Line

Consider the cases in figure 3.1. In both samples the slope coefficient is 0.90, so $1 more in parental income is associated with $0.90 more in child income. But in panel A, the data are scattered widely around the line. There are several cases in which the actual child income is more than $50,000 higher or lower than what the best-fit line predicts. The line in panel B, in contrast, has the data tightly fit around it, never straying more than $20,000 from the line and usually just a few thousand dollars from it.

To summarize, the slope coefficient captures the magnitude of the relationship: $0.90 or $0.09? The correlation coefficient captures fit: does child income always move in step with parental income? Both magnitude and fit matter, so we would like to enhance the correlation coefficient to capture magnitude or enhance the slope coefficient to capture fit. There is no good way to complement the correlation coefficient to measure magnitude, but there is a way to complement the slope coefficient to measure fit.

THE CORRELATION COEFFICIENT AND R-SQUARED

The correlation coefficient has a sibling statistic, the R-squared, that also measures fit and complements the slope coefficient nicely. Recall that the correlation coefficient (in absolute value) gives the variation shared by two variables as a fraction of their combined total variation, for example, the variation that parental income and child income share, divided by their total variation. The R-squared is also a fraction, but it has a different numerator and denominator. It gives the variation of child income predicted by parental income as a fraction of the total variation in child income. More precisely, it is as follows:

$$R\text{-squared} = \frac{\sum_{i=1}^{n}(\widehat{CI_i} - \overline{CI})^2}{\sum_{i=1}^{n}(CI_i - \overline{CI})^2},$$

where predicted child income ($\widehat{CI_i}$) is based off the estimated best-fit line. It comes from using the estimated slope coefficient (and intercept, which is easy to calculate) to predict each child's household income as an adult based on the child's parental income.[13]

If a child's actual income and the income predicted by parental income are generally close, the fraction will be closer to 1, which is the maximum value that the R-squared can take. In a scatterplot, the points (pairs of child and parental income) will generally hug the best-fit line, as they do in figure 3.1B. If the data are wildly dispersed around the line, the R-squared will be close to 0, which is its minimum value.

Like the correlation coefficient, the R-squared tells us nothing about how much child income changes for a given change in parental income. Instead, the R-squared is often said to measure the percentage of variation in Y explained by X. This is true only in the shallow statistical sense of the word *explain*, such as "umbrella usage explains 94 percent of the variation in rainfall." A better description for R-squared is that it measures the variation in one variable predicted by another regardless of whether there is a causal relationship between the two. And sometimes we don't need to explain, just predict, such as when we are planning next year's budget or knowing at what time most 911 calls are likely to occur.

The R-squared complements the slope coefficient, with one telling us about magnitude and the other about fit. If the slope coefficient tells us that each $1,000 more in parental income is associated with $600 more in child income, the R-squared tells us that parental income predicts 15 percent of the variation in child income.

THE STANDARD ERROR OF THE SLOPE COEFFICIENT

The slope coefficient has another complementary statistic related to the fit of a line. Just as we calculated the standard error of the mean, we can calculate the standard error of the slope coefficient. For the standard error of the mean, we did the thought experiment of drawing many samples of a given size, calculating the mean for each, and measuring the dispersion of sample means, which we estimated using the standard error of the mean. We can do the same for the slope coefficient: draw many samples in which each person has a parental income and a child income, calculate the slope coefficient between parental income and child income, and measure the dispersion of the slope coefficients. Mathematically, the estimated standard error of the slope coefficient between parental income and child income is

$$\text{Est. standard error of the slope coefficient} = \sqrt{\frac{\sum_{i=1}^{n}(CI_i - \widehat{CI}_i)^2}{(n-2)\sum_{i=1}^{n}(PI_i - \overline{PI})^2}}.$$

The difference between the child's actual income and the predicted income ($CI_i - \widehat{CI}_i$) is the fit of the line that we seek to measure. Less fit (more dispersion around the line) increases the standard error of the slope coefficient. Whether $CI_i - \widehat{CI}_i$ is positive or negative, squaring it makes the value positive and larger, which increases the numerator and the whole expression.

The slope coefficient and its standard error provide what the correlation coefficient or the slope coefficient by themselves could not—a measure of magnitude combined with a measure of fit. A larger slope coefficient means a steeper line (larger changes in child income for a $1 change in parental income), and a small standard error implies that the line fits the data well.

Fit is important because it affects precision. This looks like predicted values of child income tightly clustered around their actual values. Poor fit means that predictions of child income based on parental income are likely off by a lot for any given person. Fit also affects how accurately the sample best-fit line reflects the population best-fit line. The coefficient estimated from a particular sample might be very different from the population coefficient because we drew a particular sample that poorly reflected the population as a whole. The more that our sample data are dispersed around the best-fit line, the more likely that our sample slope coefficient could differ considerably from the population value.

We previously used the standard error of the mean to calculate a confidence interval for the population mean. We can do the same for the slope coefficient, with the population slope coefficient being the one that would result if we had parental and child income for the entire population. Recall that as the sample size grows, the sampling distribution of the mean becomes approximately normally distributed. The same is true for the distribution of slope coefficients from many randomly drawn samples. Thus, with a sufficiently large sample we can use critical values from the normal distribution to construct a confidence interval for the slope coefficient. Recall the basic confidence interval formula:

Mean ± standard error · critical value.

Now we replace the mean with the slope coefficient and the standard error of the mean with the standard error of the slope coefficient. The 90 percent confidence interval for an estimated slope coefficient is then as follows:

90% confidence interval for $\beta = \hat{\beta} \pm$ standard error · 1.65.

Now, when describing the relationship between parental income and child income, we can say something like "On average, $1 more in parental household income was associated with $0.90 more household income of the child as an adult, with a 90 percent confidence interval of 0.72 to 1.08" (the case where the standard error equals 0.11). This confidence interval speaks most directly to uncertainty over the population slope coefficient. As before, your audience may have little time for standard errors or confidence intervals. You should know them nonetheless and think hard about propagating estimates whose confidence intervals are large.

4

Know What It Means to Account for Potholes

The previous chapter introduced the best-fit line to summarize the linear relationship between two variables. The general method for estimating relationships between variables is called regression. Most quantitative methods classes in policy-related fields teach it in detail, making you think that the policy aide probably spends most of her time estimating regressions. In practice, the closer the aide is to those making policy decisions, the fewer regressions she'll estimate or see. Few policy makers want to see a table with regression statistics. Doing so will likely elicit blank stares or, worse, will cause key people to leave the room for more productive use of their attention.

Regressions, however, can be helpful in policy settings, although not in the way that academics typically present them. Policy makers will often value and be helped by statements like the following:

- Students with higher SAT scores graduated from college at a higher rate after accounting for their high school grade-point average.
- Rural counties received fewer grants than urban organizations even after accounting for the scores assigned to their grant applications.
- Black and White loan applicants had different approval rates even after accounting for income and credit score.
- Lower-income neighborhoods received less road spending than higher-income neighborhoods after accounting for the number of potholes in each.

The policy aide should have an intuitive and precise understanding of what it means to account for X, which may be age, income, race, or potholes. Comparisons that account for X—or to use the more academic phrase, control for X—aid in diagnosing problems, debunking specious claims, and understanding how programs are working. Accounting for X is important because one of the most common statistics used in policy discussions are differences in simple means: rural areas compared to urban areas, Black populations to White populations, participants to nonparticipants, and so on. Though easy to understand, simple differences can mislead. Accounting for obvious correlates can greatly improve comparisons.

To illustrate, consider a scenario in which residents of a city have complained about unfairness in the distribution of road-repair spending. Some claim that decent roads in higher-income neighborhoods receive more spending than bad roads in lower-income neighborhoods. The mayor asks the Department of Public Works to investigate, and you, as a policy aide in the department, receive the task. You download neighborhood income data, and from the department's own data, you calculate the number of potholes per mile for each neighborhood at the beginning of the fiscal year and the road spending per mile over that fiscal year.

After merging the department data with the neighborhood income data, you use your favorite statistical package to estimate a regression, where road spending per mile (in \$1,000s) is the dependent (Y) variable and potholes per mile and median household income (\$1,000s) are the independent (X) variables. The regression yields the following precisely estimated numbers:

Road spending per mile = −62 + 1.9 potholes per mile

+ 1.2 household income.

The equation implies that one more pothole per mile is associated with \$1,900 in greater road spending per mile, holding constant neighborhood income. And \$1,000 in higher household income is associated with \$1,200 in greater spending per mile, holding constant potholes per mile.

Few people will receive that equation with enthusiasm. Nonetheless, it gives what we need to address the concern that much spending

Figure 4.1 Road spending in higher- and lower-income neighborhoods

is related to neighborhood affluence and not road quality. With the estimates, we can put numbers to the following thought experiment: if the average higher-income neighborhood had the same number of potholes per mile as the average lower-income neighborhood, how would spending differ between the two?

Splitting neighborhoods into the lower-income half and the higher-income half results in two groups with average median incomes of $55,000 and $72,000. The coefficient on income permits translating the difference in income (= $72,000 – $55,000 = $17,000) into a difference in spending (= $17,000 · 1.2 = 20,400). Thus, the average higher-income neighborhood receives about $20,000 more in road spending after accounting for road quality.

To put the extra spending in context, spending in the lower-income neighborhood with the average number of potholes per mile (ten) is $23,000. (This can be found by plugging ten potholes and $55,000 in income into the regression equation.). By comparison the average higher-income neighborhood with the same number of potholes has $43,400 in road spending per mile. So, once we account for differences in potholes, higher-income neighborhoods receive almost twice as much road spending on average than lower-income neighborhoods! Figure 4.1 shows the story.

The first pair of bars in the figure—the simple means—is what someone will point to as evidence that there is no problem: higher-

and lower-income neighborhoods receive the same spending per mile. The comparison, of course, ignores differences in road quality. Once accounted for through regression, the disparity emerges.

Understanding the Mechanics and Essence of Accounting for Potholes

To forge a precise and intuitive understanding of how exactly regression accounts for potholes, we must break it into two steps. The regression accounts for potholes by first purging the variation in income that is linearly related to potholes. To see how the purging works, consider the best-fit line relating income to potholes, as shown in figure 4.2.

The best-fit line represents the income-pothole relationship. The difference between actual income and income predicted by the line, what is called residual income, is what we want to use to estimate the spending-income relationship. In the second step, then, we estimate the best-fit line between spending and *residual* income, which is unrelated to potholes.

Does the resulting coefficient truly give the spending-income relationship while accounting for potholes? Not in the sense that we actually find lower-income neighborhoods and match them to

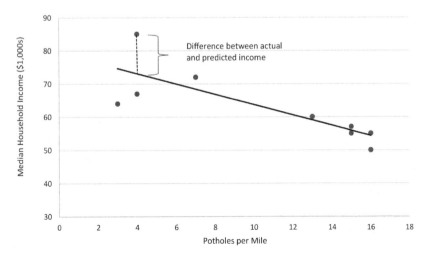

Figure 4.2 Variation in neighborhood income unrelated to potholes

higher-income neighborhoods with exactly the same number of potholes. (Some methods actually seek to do this.) What is needed for the "accounting for" or "holds constant" interpretation is not that every lower-income neighborhood has an exact match, but that the two groups have the same number of potholes on average.

Having two groups be similar on average is what a randomized experiment seeks to do (often referred to as a randomized controlled trial; see chapter 6). Consider the strange experiment in which we randomly distribute potholes across neighborhoods without regard to income or any other characteristic. Because of the randomization, we expect (and can verify) that higher- and lower-income neighborhoods have the same number of potholes on average. In this sense, the comparison keeps the number of potholes constant across the two groups.

The regression mimics the experiment by creating a version of the neighborhood income variable (residual income) stripped of its relationship with potholes. Because of how the best-fit line is defined, neighborhoods with higher residual income have the same number of potholes as neighborhoods with lower residual income on average, just as when potholes were randomly assigned to neighborhoods and therefore unrelated to income. In this way, the best-fit line between spending and residual income gives the spending-income relationship after accounting for potholes.

Regression Pitfalls

Looking for Nails Because We Have a Hammer

With its ability to account for road quality and age and other variables, regression can be a powerful statistical tool for the policy aide. Yet it is often overused and clumsily wielded. Too often we regression people are like carpenters who love hammers and therefore see nail-like things everywhere. Naturally, a professor teaching statistical analysis for policy often asks students to identify a project in which they can practice the regression methods taught in class. Outside the classroom, we should flip the approach, starting with a problem or question and then selecting the appropriate statistical tool. This is the same idea as first considering our purpose before concluding whether data are useful (chapter 2). In many instances, simple statistics thoughtfully applied yield more insight than throwing variables into a regression and looking at coefficients (or worse, the *t*-statistics and *p*-values discussed in chapter 5).

Think about what you seek to accomplish. A good rule of thumb is to not estimate a regression without a clear purpose. Wanting to see what the data look like is not a clear purpose. Suppose you are exploring patterns in grant awards to schools. Don't just dump variables into a regression. Think about the problems you seek to detect. Do you want to know if grants are going to only certain schools or if they are used for purposes inconsistent with the program's rules? Define your purpose, form a working hypothesis, and let it guide what you put into the regression and whether you use regression at all.

Not Understanding Variable Units, Definitions, and Coefficients

Anyone estimating a regression should be able to use the estimated coefficient in a sentence that makes sense to the typical newspaper reader. The task requires understanding variable definitions and the practical significance of their units. Recall from chapter 2 that knowing units is more than knowing that income is measured in Mexican pesos. It is knowing whether a peso buys ten beers or one-tenth of a beer. In a rush to estimate a regression, we often gloss over such basics.

A shallow understanding of units and definitions often makes an unflattering appearance when we talk about our coefficient estimates. We might say—as many of my students have—that a 1-unit increase in potholes is associated with a 1.9-unit increase in road spending. A dazed look then comes over the faces in the audience. If we can recapture their focus, we should instead say, "An additional pothole is associated with $1,900 more in road spending in the neighborhood, either for repairing potholes, resurfacing the street, or repaving it altogether."

See how knowledge of definitions gives a more precise meaning to the coefficient estimate? Road spending could mean many things. In an analysis of potholes and road spending, the audience could think that spending includes only spending on filling potholes, one by one, and does not include spending on total repaving, which is just a more comprehensive and longer-term approach to filling potholes.

Estimating Unintuitive Coefficients

Another regression pitfall related to interpretation is defining variables that make the coefficients hard to interpret. Several examples illustrate the importance of defining variables and specifying regressions so as to make their estimates easy to interpret and sensible once understood.

The road-spending regression gave the estimate that each pothole per mile is associated with $1,900 more in road spending per mile. We can simplify the statement further and just say that each pothole is associated with $1,900 more in road spending. To see this, multiply the pothole regression equation by the variable miles of road. The 1.9 coefficient is now next to the variable potholes, and the dependent variable is simply road spending.[1]

The average newspaper reader can understand spending $1,900 per pothole. But what if we had defined the dependent variable as spending per household in the neighborhood? Now we can't make the dollars-per-pothole statement. We can talk only of the relationship between potholes per mile and spending per household, which is cumbersome. Or, imagine defining the dependent variable as the percentage of the city's road budget going to the neighborhood. Now, the coefficient on potholes per mile gives the percentage-point increase in road spending going to a neighborhood. In terms of clarity, we've gone from foggy to opaque.

What if we had instead defined the dependent variable as the natural log of road spending per mile? We're still in the dark regarding interpretability, and the coefficient has become even less sensible, even to people who eat regressions for breakfast. The coefficient on potholes per mile (×100) is now the percentage increase in spending for each additional pothole per mile. Come again? The coefficient is not just hard to understand; it also makes little sense. It means that one more pothole per mile brings the same proportional change in spending regardless of whether a neighborhood initially had $100 dollars in spending per mile or $10,000 in spending per mile, which would translate into very different dollar amounts. This is hard to believe if we think that potholes generally cost about the same to address regardless of where they are.

Thinking the Work Is Done Once We've Estimated Coefficients

Another regression pitfall is thinking that the work is done once we have estimated the coefficients. Much work remains, most notably the work of squeezing meaning from the regression statistics, which furthers the purpose that drove the regression initially. Imagine that you present to the head of the Department of Public Works the regression showing that each $1,000 more in household income in a neighborhood is associated with $1,200 more in spending per mile. Put yourself in his shoes and think about what you would do with that number. Even if you had a PhD in regression, it would be unclear how the number speaks to the concerns that reached the mayor's office. More work is needed.

The effort to squeeze meaning from regression statistics must be guided by your purpose for estimating them initially. In this case, it was to see whether the city unfairly distributes road spending, favoring higher-income neighborhoods whose roads are in decent condition. Reporting the coefficient estimates to the head of the Department of Public Works or saying that income and potholes are correlated with spending is like building a bridge partway across the river and going home. We must finish the bridge and connect the shore of statistics with the shore of the policy maker's need.

In the pothole example, we used the coefficient on potholes to perform the following thought experiment: if a higher- and a lower-income neighborhood had the same number of potholes per mile, how much more would the city spend on the roads of the higher-income neighborhood? The answer, made possible by the regression, is that the average higher-income neighborhood would receive almost twice as much road spending as the average lower-income neighborhood, if both had roads of the same condition. This speaks directly to the concerns raised. The head of public works will understand it. The mayor will understand it. We've built the bridge. Whether or how they want to address the issue is another matter, but they will go forth understanding that current policy directs more road spending to higher-income neighborhoods.

Squeezing meaning from regression coefficients also involves assessing their magnitude and carefully selecting the right adjectives to communicate the result of our assessment. Is the estimate large in some sense? In what sense? See chapter 5 for distinguishing large and small coefficients and explaining the difference.

Practical Uses of Regression

Accounting for X

A common role of regression in policy is what I have already highlighted—to produce statistics accounting for X, be it race or age or temperature. In a way, regression is always accounting for other variables, but sometimes it is used more narrowly to account for differences in obviously important characteristics and allow for a clearer and more helpful comparison of simple statistics. The fictitious road-spending example fits the category. It enabled a comparison of spending across higher- and lower-income neighborhoods while accounting for road quality, an obvious determinant of road spending. Accounting for X is often done using the language of adjustment. We could have spoken of the relationship between neighborhood incomes and road spending *adjusted* for road quality.

One can find numerous real-world examples of adjusted statistics. The White House Council of Economic Advisers regularly releases data-rich policy reports meant for a broad audience, including congressional staffers, journalists, and think tanks. Rarely will you find a table of regression statistics in a council report. You will, however, find behind-the-scenes uses of regression, like in its 2019 report *The State of Homelessness in America*. The report shows a scatterplot with the temperature-adjusted rate of homelessness in different US regions on the y-axis and a measure of housing affordability on the x-axis.[2]

The scatterplot would be hard to interpret without adjusting for temperature because counts of the homeless population usually occur on a single night in late January. Clearly, we expect fewer people to be sleeping on the street in January in North Dakota than in Southern California. Thus, much of the variation in homelessness is driven by stark differences in climate, not in economics or policy. Policy makers cannot turn North Dakota into Southern California, but they can affect incentives surrounding housing and homelessness. Stripping out the variation in homelessness related to climate helps focus the reader's attention on affordability and non-climate-related variation in homelessness.

Another common example of adjusted-for-X uses is to create trend figures adjusted for time-varying characteristics. The share of the working-age population working or seeking work is called the labor-force participation rate. It varies over time in part because the age profile of the population changes and because older people have a lower propensity to seek work. Regression can permit the creation of a labor-force participation rate adjusted for the share of the population in different age ranges. Just like we wanted to see road spending for neighborhoods with different incomes but roads of similar quality, we might want to compare the labor-force participation rate in different years but with populations of similar age.

Identifying Patterns

A related use of regression is to reveal patterns, such as how a program is working and who is benefiting. I once oversaw a team of graduate students who sought to help a state agency understand which types of schools tended to receive its grants. One option to determine this would be to divide schools into two groups, recipients and nonrecipients, and compare the means of various variables across the two groups. Such a comparison might reveal that recipient schools are more often predominately White and have more administrative staff per student. The simple mean comparison, however, does not help us assess whether schools with more administrative staff receive more grants regardless of the racial composition of the student body (or vice versa). Put differently, the simple comparison does not tell us if administrative staff matters once we've accounted for race. As the pothole example showed, a regression with multiple variables does this.

A similar example of this use of regression is the federal Department of Housing and Urban Development's effort to understand patterns in mortgage lending as part of its charge to enforce the Fair Housing Act. Its Office of Policy Development and Research studied patterns in high-interest lending across communities, relating the share of high-interest loans in a community to characteristics such as share of the population in different racial or ethnic groups, income groups, education groups, and so on.[3] Other examples abound, such as those looking at patterns in who has health insurance, who takes advantage of energy-efficiency tax credits and rebates, and which communities have air-quality monitors.[4]

Estimating Causal Effects

Another common use of regression is to estimate the causal effect of something—health insurance coverage on emergency department use, job training on earnings, and so on. The ability of regression to account for confounding variables makes it a natural tool for developing causal estimates. It helps us run the following thought experiment: if we compare a worker in two states of the world, one in which he participates in job training and one in which he doesn't, how would his post-training earnings differ? The comparisons accounting for X are in this spirit and are meant to improve comparability so as to suggest a causal effect as opposed to a mere spurious correlation. Such relatively simple comparisons, however, are usually not forcefully argued as isolating causal effects. An analysis seeking to make causal claims often requires more work and is the focus of chapter 6.

5

Know Large from Small and Explain the Difference

My statistics education did not prepare me for one of the most important skills for using statistics for policy—distinguishing large from small. Statistics textbooks and classes center on teaching statistical concepts, formulas, and calculations. They do not equip us to assess magnitude, that is, whether a statistic is large or small, alarming or ignorable. The lack of emphasis on assessing magnitude is understandable. How many professors can confidently teach statistics and ethics, mathematics and rhetoric? Too rare are statistics educators like Robert Abelson who taught that "the purpose of statistics is to organize a useful argument from quantitative evidence using a form of principled rhetoric" (by *principled* he means rhetoric that fits the evidence).[1]

People get PhDs in statistics or rhetoric, not both, and they teach according to their training and strengths. Most statistics educators are at home with formulas and their properties. They can provide the proof for the law of large numbers but can offer little guidance on which adjectives to put next to the (hypothetical) finding that 1 in 1,000 natural-gas wells contaminates drinking water supplies. The vacuum of guidance on magnitude is unfortunately filled by a formula and a rule that deems a number statistically significant or not. The formula and rule help us assess magnitude as much as a person's height tells us her greatness as a volleyball player. Height is relevant, but it is neither necessary nor sufficient for greatness on the court.

Before discussing the strengths and shortcomings of statistical significance, first understand that assessing magnitude is both controversial and important.

Large and *Small* Are Controversial Words

Hydraulic fracturing (fracking) is a method of injecting fluids into rock to extract oil or natural gas trapped within it. It is controversial in part because of concerns that it can contaminate the drinking water of nearby residents. In 2009, a committee of the US House of Representatives asked the Environmental Protection Agency to study the relationship between fracking and drinking water.[2] In June 2015, the agency released a draft report of its findings. Summarizing nearly a thousand pages that reviewed the scientific literature and independent data analysis, the draft report concluded, "We did not find evidence that [fracking-related activities] have led to widespread, systemic impacts on drinking water resources in the United States . . . the number of identified cases where drinking water resources were impacted are small relative to the number of hydraulically fractured wells."[3]

The conclusion was based in part on data from eleven states over six years during which tens of thousands of wells were fracked. Over that time and for states with data, the Environmental Protection Agency (EPA) identified 376 fracking-related spills that occurred, 18 of which reached the groundwater.[4] The report also recognized that local water-quality data from before and after fracking is rare relative to the number of fracked wells, making it hard to accurately quantify the frequency of contamination or attribute it to fracking.

When released, the draft report drew harsh criticism from environmental groups and praise from industry. Of the EPA's thirty-member science advisory board, twenty-six supported a critique saying that the conclusion of no widespread, systematic impacts was not quantitatively supported, nor were the terms *systematic* or *widespread* defined. Four other members said that the language needed no modification.[5]

The conclusions of the draft report were then cleansed of adjectives. The final report, released in December 2016, omitted the terms *widespread, systematic,* and *small.* After years of independent research and reviewing 1,200 sources of data and information, the agency refrained from commenting on how much risk fracking poses to drinking water. Its sterilized conclusion centered on naming the factors that are likely more important than others in determining the risk of contamination. The strongest part of the conclusion was "Cases of impacts were identified for all stages of the hydraulic fracturing water cycle."[6]

At the core of the controversy surrounding the report was not the manipulation of data, the suppression of data, the omission of data, or the statistical decisions made to summarize the data. The controversy was about the adjectives attached to the statistics. Some may argue that the EPA had too little data to calculate the statistics that would have supported stronger conclusions. Perhaps. But even if the EPA had perfect data showing a 0.1 percent chance of water contamination (1 in 1,000 wells), it would have received harsh critique from somewhere for whichever adjective it used to describe the risk.

Not only did the report avoid controversial words like *small*. It even avoided equipping readers with numbers to aid them in making their own assessment of magnitude: the EPA reported that "cases of impacts were identified," but the fifty-page executive summary omits a total count of identified cases of impact.

Small and *large* are controversial words. It's that simple. People and organizations who want to avoid controversy cleanse them from their reports. The effect: their sterilized conclusions leave readers feeling like they just ate air. Yet for policy-oriented people, assessments about large and small are as important for aiding policy makers as data are for calculating statistics.

Distinguishing Large from Small Is Important

Using adjectives like *small* and *large* might stir controversy, but having them—and reasons for them—is important if our findings are to be intelligible and helpful. The policy maker needs adjectives and reasons for them. He can reject our adjectives and choose others, but we are not serving him well by providing numbers without adjectives and reasons.

Suppose that I drive into the auto repair shop and ask the mechanic if I should buy new tires, and the mechanic examines the tires and reports that they have 6 millimeters of tread left. Then she looks at me in silence as if her response just answered my question. I repeat the question, "Should I buy new tires." Her response: "Oh, that's your decision. I just measure the tread." Really? Help me out. I don't look at tires all day, so 6 millimeters means nothing to me. Do tires start with 8 millimeters of tread or with 18? Do they start to blow out at 6 millimeters or 3 millimeters? I need contextual information relevant to assessing the magnitude of 6 millimeters.

If the mechanic insists on being a measurement-only type, I will make a poorly informed decision about new tires or visit another repair shop. There is no avoiding a decision. I will buy new tires or keep the old ones. And by the way, I need to pick up my kids at 4 p.m., so I need to decide soon.

Like the mechanic, you can say: "I just take measurements. Others decide what to do with them." And if you want to just do tire statistics for tire statistic yearbooks, that's fine. But if you want to serve people well and repeatedly, they need more than your measurement; they need your assessment of what you have measured. I am not saying that the policy maker should defer to your assessment; he must own the decision since he will bear responsibility for its consequences (just like I would have to pay for new tires or bear the risk that comes with keeping the old ones). Rather, your assessment and its supporting reasons will aid the policy maker in making the decision.

The EPA fracking report illustrates that reporting information without a well-reasoned, transparent assessment of magnitude will be unhelpful to policy makers. Consider the report's conclusions. One part is that factors that are "more likely than others to result in more frequent and more severe impacts" include "spills . . . that result in large volumes or high concentrations of chemicals reaching groundwater resources." Obvious and unhelpful. Or take the conclusion that "cases of impacts were identified for all stages of the hydraulic fracturing water cycle." Also unhelpful. By 2016, more than seventy thousand horizontal wells had been drilled in the United States (horizontal wells are almost always hydraulically fractured wells).[7] It would be astounding if there were zero documented cases of contamination. The report therefore says little about the risk that nearby residents face or the adequacy of current regulation.

The EPA report is a prime example of providing information without the contextual information and reasoning that aid in assessing magnitude. There is no sifting and weighing of the numbers in a way that suggests the appropriateness of one policy over another. The report does little to inform questions like "Should we regulate more now?" Acknowledging the perpetual need for more data and newer data is not an excuse for avoiding frank assessments of magnitude. Today we decide if more regulation is needed. By not highlighting the magnitude of the risk suggested by the data and studies, the report authors removed themselves and their report from the policy conversation. Parties on both sides of the issue would find plenty of material in the report to be used to throw at the other side, but the report itself would do little to change anyone's thinking on which policy makes sense. This is why assessing magnitude is important—it makes the findings intelligible and relevant for conversations about what to do. The policy maker will make a decision informed by your assessment or go elsewhere for people prepared to assess magnitude and put adjectives next to numbers.

To be clear, the task is more than slapping adjectives on the page. In a moment anyone can drop in *small* or *only*. It is not the adjectives themselves that help the policy maker but the reasons for the choice of *large* or *small*, of *systematic* or *isolated*. In fact, the advisory board criticized the EPA's draft report because there was no explanation of why the EPA selected certain adjectives. Adjectives without reasons give the appearance of bias, an unthought-through predisposition to see something as good or bad, large or small. It also can be condescending toward the audience. Stating that the risk of harm is *only* 0.1 percent without explaining the reason for the *only* assumes that the magnitude is obvious to all, which it never is, or that the audience should not critically engage but instead trust the assessment of the expert.

It is easy to report measurements as if they were answers to key questions, as if the mechanic's report of 6 millimeters of tread on my tires answered my question about the need to buy new ones. My fictional example can be readily replaced with real ones, where someone asks about magnitude and its implications for decisions, and the expert just reports a measurement. As the COVID-19 Delta variant was spreading through the United States in 2021, a news reporter for a national media organization asked a public health expert if we, the public, should be concerned about the rise of the variant. The expert responded by saying that the number of hospitalizations of young people were hitting record levels. At first glance, the response seems to answer the question. Record-breaking hospitalizations sound concerning, so yes, we should be concerned about the Delta variant.

Record numbers, however, are not always important numbers: the prior record might have been insignificant, and a number slightly larger than a small number is still small despite breaking records. My risk of a heart attack today is probably the highest it has ever been in my life—I've never been older—but that does not mean that I should be concerned or change my behavior.

To answer the reporter's questions, we need context and reasoning about magnitude. What is the actual number of hospitalizations? The prospects for even higher numbers? The fatality rate among those hospitalized? More broadly, how does the risk of dying from the Delta variant compare to other risks that young people face? A good response to the reporter pulls together such numbers and makes an argument about large or small, high risk or low risk. In short, a good answer actually answers the question, Should we be concerned?

Distinguishing Large from Small Is Hard

Choosing the right adjectives to describe a number is hard and cannot be reduced to a formula or a rule of thumb. A few weeks into the semester, freshmen statistics students can grab the data and calculate risk, but assessing the seriousness of 1 in 1,000 wells contaminating drinking water requires much more. It is difficult work, and statistical significance inadequately fills the magnitude vacuum left by most statistical education.

Consider again the hypothetical example where an abundance of good data permit calculating that 1 in 1,000 fracked wells (0.1 percent) leads to a drinking water contamination event. You toiled to assemble the right data and ensure its quality. You are confident in the number. You share it with a friend at the end of the data-crunching day who responds: "Is that a big deal? Seems rare." Is it a big deal? Does fracking pose a high risk of contamination? There is no adjective generator that can suck in the 0.1 percent finding and spit out the right adjective. None. Choosing adjectives requires that we dig our hands into the soil of the subject.

For the fracking example, describing the 0.1 percent number might involve contextual digging into the following topics:

- The particular contaminants and the health consequences of the contamination events. Are the events killing people? Putting them in the hospital?
- The persistence of the contamination and the economic cost of remediating it. Does contamination naturally flush out of the system in a week, or does it persist indefinitely unless millions of dollars are spent on cleanup?
- The source of the contamination. Are one or two companies responsible for all contamination events, suggesting that the risk is near zero in most cases but high for the wells of particular companies?
- The contamination frequency of other activities. Do 1 in 100 hog-manure pits lead to drinking water contamination, or is it 1 in 10,000? This gives a sense of society's risk tolerance in other domains.

As the example shows, assessing magnitude requires more effort to grasp the real-world consequences of a contamination event. This takes time, thought, and data. You thought the work was done when you calculated the 0.1 percent number. But if you want to assess the magnitude of the statistic and actually further people's understanding of the issue, much work remains. Each of the bullet points above could be its own research project!

Given the hard work and subjective nature of assessing magnitude, it is not surprising that many people instead turn to a formulaic and seemingly objective method, the method of statistical hypothesis testing, or what I will simply call statistical significance. A *statistically* significant effect or difference is often implied to be a *practically* significant effect or difference. But one does not imply the other. To see this, we must understand what it means to determine that an effect is statistically significant.

The Basics of Statistical Significance

Imagine that media reports suggest that people do not want to live near the city's trash incinerator and that homes near it are worth less than homes farther away from it. Are homes near it actually worth less than comparable homes further away (putting aside the question of causality addressed in chapter 6)? Common practice for assessing whether a difference is real is to see if it is statistically different from 0. If it is, it is labeled a statistically significant difference, which is often interpreted to mean a difference that a typical person would care about.

So, we collect data on recently sold homes and divide them into a near-incinerator group and a far-incinerator group and calculate the mean sales value for each. Near-incinerator homes have an average value of $300,000 compared to the $330,000 value for far-incinerator homes. The statistics are consistent with the media reports and suggest lower values near the incinerator. Yet they alone should not give us great confidence that the difference is real. By "real," I mean something that would be observed if we had data on the entire population of homes in the near and far areas. Recently sold homes might only account for 5 percent of all homes in the near and far areas. If a different mix of homes had been sold in the two areas, we might observe very different sample means. And if we could imagine every home being sold, the average price in the two areas could be nearly the same. In this sense, the difference between the first samples would not have been real.

The notion of statistical significance emerged precisely because two samples will probably have different means even if there is no difference in the population. In the incinerator example, this is where homes in the near and far areas actually have the same values. The $30,000 difference might merely reflect the sample we drew, not that the two areas actually have different average housing values.

The method for assessing whether the difference in sample means reflects a real difference in population means begins by adopting the hypothesis that there is in fact no difference in population means. We assume, for example, that the $30,000 difference in sample means merely reflects sampling error—by the luck of the draw, we drew houses that were imperfectly representative of near and far areas.

The next step is to calculate a measure of the consistency between the hypothesis (no actual difference in means) and the sample data. The calculation depends on the difference in sample means and the estimate of the standard error of the difference. The standard error measures how much the sample difference in means varies on average from the true population difference in means. It is the same standard error discussed in chapter 3, only here we are working with the standard error of a difference in group means as opposed to the mean itself. Its formula is therefore a modification of the basic formula for the standard error of the mean discussed in chapter 3:

$$\text{Est. standard error of a difference in group means} = \sqrt{\frac{\text{Est. pop. variance}_1}{\text{Group size}_1} + \frac{\text{Est. pop. variance}_2}{\text{Group size}_2}}.$$

When calculating the standard error of the mean, we took the square root of the variance divided by the sample size. For the standard error of the difference in group means, we take the square root of the sum of the respective group variances divided by the group sizes.

The ratio of the difference in group means over the standard error is called the t-statistic:

$$t\text{-statistic for a difference in means} = \frac{\text{Group mean}_1 - \text{Group mean}_2}{\text{Standard error of the difference}}.$$

Under the hypothesis that the population difference in group means is 0, the t-statistic has an approximate standard normal distribution (in large samples) with a mean of 0 and a standard deviation of 1. This means that if the hypothesis is true, we will generally see a t-statistic close to 0 since the numerator (the difference in group means) will be near 0. Rarely will we see the t-statistic greater than 1.65 or less than −1.65.

Because we know the distribution of the statistic under our assumptions, we can assign a probability to observing extreme values for the t-statistic. Only 10 percent of the time will we observe a value more extreme than ±1.65, and only 5 percent of the time will we observe a value more extreme than ±1.96. Common practice is to reject a hypothesis whose t-statistic is beyond ±1.96, which is known as the 5 percent significance level because such extreme values are observed less than 5 percent of the time under the hypothesis. If our sample home values and standard error give an extreme t-statistic, we reject the hypothesis that the near and far home values in the population are the same and deem the difference to be statistically significant.

Sometimes people skip the *t*-statistic and simply report a measure of its extremeness, the *p*-value. The *p*-value gives the probability of observing such an extreme (or more extreme) *t*-statistic in a world where the hypothesis is true and the *t*-statistic is normally distributed. It therefore ranges from 0 to 1, with lower values indicating more divergence between the hypothesis and the sample data. A *p*-value of 0.03 means that if the hypothesis of equality of group means were true, we would observe such an extreme *t*-statistic 3 percent of the time. In other words, the hypothesis is not very consistent with the sample data.

One common rule is that a *t*-statistic in the bottom 2.5 percent of the tail or the top 2.5 percent is an indication that the difference in group means is statistically different from 0 or simply "statistically significant." Otherwise, the means are deemed to be not statistically different. The rule reflects a particular weighting of errors (statistical loss function). It is greatly concerned about treating an unreal difference in means as real, what is known as a type 1 error, or a false positive. In contrast, it is far less concerned about treating a real difference as unreal, a type 2 error, or a false negative. Ironically, the rule for deeming a difference significant has led many to declare a practically small difference as significant, that is, the treating of something essentially unreal as real. See the next section for more on the conflation of statistical significance with practical significance.

Note that *t*-statistics can take many forms beyond that of a difference in group means. A common use is to assess whether a slope coefficient relating one variable to another is statistically different from 0. For example, is the relationship between housing values and distance from the city's incinerator statistically significant? The hypothesis can also vary; we could hypothesize that the difference in group means in the population is less than $10,000 rather than being 0. The most common hypothesis, however, is that there is no difference in group means. Any observed difference reflects sampling error.

A help-wanted sign for significance testers might read "Apply Today—No Knowledge or Prior Experience with the Industry Needed." Drawing a conclusion about the statistical significance of a number, in this case the difference in group means, requires no knowledge of the subject surrounding the data. One needs only to know how to perform calculations or, with today's software packages, how to properly organize the data and press the right buttons. If homes near natural-gas wells have an indoor radon level that is 1 picocurie per liter higher than homes far from wells, one can quickly label the difference statistically significant or not without any understanding of whether an additional 1 picocurie per liter will kill or if it is the same increase as one would get from using the microwave.

In contrast, assessing practical significance (magnitude) requires understanding the subject under study. What is a picocurie, and what does radon do to the body? Is an extra picocurie per liter a big deal? Does it matter if most homes are going from 1 picocurie to 2 picocuries, or if they are going from 100 picocuries to 101 picocuries? Calculating the difference in group means is the beginning of the hard work of assessing whether natural-gas wells pose an important radon risk to nearby residents. Statistical significance is commonly used as a substitute for this hard work.

Tests of Statistical Significance Cannot Tell Large from Small

Producers of statistics want to talk about significance and magnitude. Why else would we tediously compile and clean data? We want to discover something important, a number so large it cannot be ignored or so small that we should stop talking about it. Consumers of statistics also want to hear about magnitude. They want to know the "so what" of a number. Yet many of us producers of statistics also want to avoid controversy and charges of being subjective or biased in assessing magnitude.

Into this conundrum enters the statistical test for significance, a mechanical procedure and mathematical rule that permits putting the word *significant* next to an estimate of an effect while giving off an aura of mathematical rigor and scientific authority. Widespread reliance on statistical significance for assessing magnitude is not just because statistics classes emphasize it. Consumers and producers of statistics want what it appears to provide—its straightforward interpretation (the effect is significant!) and its veneer of objectivity (the formula, not the person, determines significance).

The focus on statistical significance in the social and natural sciences is of epidemic scale. Abundant and accessible data combined with easy-to-use software packages allow someone with one statistics class and no subject-matter knowledge to begin to report so-called significant mean differences between Black and White populations, participants and nonparticipants, and homes near and far from a gas well. The widespread use of statistical significance and p-values brings widespread error, the most common being the equating of a statistically significant difference with a practically significant difference. It is using statistical significance as a crutch for not doing the hard, controversial work of assessing magnitude. Yet it is an inadequate crutch. Crutches are wooden and unable to navigate uneven ground, and so are the rules of statistical significance. Assessing magnitude requires traversing mountainous terrain. Crutches will not do. How will they not do? In their must-read chapter "Lady Justice v. Cult of Statistical Significance," Ziliak and McCloskey explain that "the core problem is that statistical significance is neither necessary nor sufficient for testing an ethical, scientific, commercial, or material fact in a court of law or of scientific and business opinion."[8] To understand the statement, recall that significance is based on the t-statistic taking on an extreme value. The t-statistic is a ratio of two numbers, a measure of the effect on the top and a measure of uncertainty on the bottom. It therefore suffers the curse of all ratios: a very large bottom can make a large top look small. Likewise, a very small bottom can make a small top look large. Statistical significance is not necessary because the effect (the top) can be large without the t-statistic being extreme (significant). And it is not sufficient because the effect can be small but the t-statistic extreme.

To illustrate that statistical significance is not necessary for an important finding, consider our gas well and drinking water example again. Suppose that we have data on the groundwater quality of 1,000 locations near wells and 1,000 locations far from them. Suppose further that ten wells (1 percent) have leaks, leading to methane entering the groundwater. The ten wells lead to 10 milligrams per liter of methane in the groundwater. All other locations—near wells or not—have an average methane concentration of 1 milligram per liter. The standard deviations of the two groups are 1.5 near wells and 1.0 far from wells. Applying the formula for a statistically significant difference in group means gives a t-statistic of 1.58.[9] With the t-statistic insufficiently extreme by standard thresholds, we would belittle the difference in means as statistically insignificant. Or we might say that wells had no statistically significant effect on methane in groundwater, which most people would hear and think, "Great, wells don't affect groundwater." But this just reflects the statistical test's limited ability to detect a true effect, an ability known as power.

Many people probably find the actual difference concerning. Very small probabilities of a car crash get people to pay for comprehensive car insurance. A 1 percent chance of a bad outcome gets most people on the phone with their insurer or, in this case, their local government representative or the state's Department of Environmental Protection. Consider the implications more broadly. If 50,000 locations across the United States have wells, the 1 percent rate of contamination implies 500 contaminated locations and maybe tens of thousands of affected people.

It is also easy to find examples in which a statistically significant difference is an ignorable difference. Recall that the standard error decreases with the sample size. With enough data, the smallest of differences will yield a statistically significant difference. In one study, a colleague and I compared the characteristics of farmers who responded to an important survey with those who ignored it. The average household size of respondents was 2.85 persons; the average of nonrespondents was 2.92, a difference of 0.07 persons. Because the sample included nearly 200,000 farmers, the 0.07 person difference was statistically significant at the 1 percent level (p-value of 0.003).[10] Only the most magnitude-oblivious person would proclaim that respondent farmers have significantly smaller households than nonrespondents.

Because of the relationship between sample size and the t-statistic, small but statistically significant differences become common as data sets grow larger. Data sets based on electronic medical records, satellite imagery, cell phone signals, and social media posts have hundreds of thousands or millions of observations. In such applications, statistical tests, especially ones built around a hypothesis of "no difference," are even less capable of distinguishing large from small.

In short, statistical tests are never substitutes for assessing magnitude. They are about one type of uncertainty—do we have the right number, or might more representative data yield a very different number? They are not about magnitude—assuming we have the right number, is it large or small, concerning or ignorable? Instead, statistical tests complement assessments of magnitude. Two numbers both assessed to be large and concerning should not be handled the same if one has a narrow confidence interval, all of which covers large values, and one has an interval half of which covers what small values. There is more cause for concern in the first case than in the second.

The Effect Size, Correlation Coefficient, and Odds Ratio Are Also Inadequate

Outside of statistical testing, a common way to gauge magnitude is to calculate what is called the effect size. Measures of effect size take different forms, but a common one is Cohen's d, which measures effect size by comparing a difference in group means, say, math test scores, to the standard deviation of test scores. It measures the effect of the education intervention relative to how much test scores vary from student to student. Cohen argued that particular values of the statistic could be deemed small, medium, or large. For example, he deemed a value of 0.20 or less as small.[11]

Dividing the change in test scores, for example, by the standard deviation has a benefit. Difficult tests are likely to have more varied scores than easy tests. Struggling students will bomb a hard test, and the kids who go to math camp during summer vacation will ace it. By dividing the change in scores by the standard deviation, the effect size can improve the comparability of changes in scores across different tests. But the benefit of dividing by the standard deviation comes at great cost. Communicating to policy makers in standard deviations is a good way to be ignored. People know what it means for 80 percent of students to pass the state test versus 50 percent of them failing it, but they will struggle with the meaning of test scores improving by 0.20 standard deviations.

The more fundamental problem with measures of magnitude like effect size is that they are applied universally. But the question of size, or what I have been calling magnitude, rests on the particulars of a case. Suppose that a school experiments with serving lunch forty-five minutes earlier. The principal suspects that students may learn better if they have lunch before being overtaken by brain-freezing hunger, which causes them to overeat and be lethargic for afternoon classes.

And suppose that the change increases math scores by 0.15 standard deviations.

According to Cohen (1992), 0.15 is a small effect. I would agree with the "small" judgment if the school had spent millions of dollars on consultants and hundreds of hours to implement a novel and supposedly high-impact teaching strategy. But in the lunch case, the only cost to the school was the time it took the secretary to modify the class schedule and communicate with the cafeteria staff. Gaining a 0.15 improvement for free is a huge win. More generally, the magnitude of an effect must be assessed in light of the effort that went into producing it, an idea explored in a few sections.

As with the effect size, some have categorized correlation coefficients as small, medium, or large. The categories may have use in academic settings where vague notions of the strength of a relationship suffice. But they will not do when deciding how to spend the scarce dollars of the city government. We need measures of how much tax revenues will increase. Knowing that there is a 0.3 correlation coefficient between spending on parking enforcement and meter revenue is useless. It does not help us think about the trade-off of putting an additional dollar here versus there. The city budget chief needs dollars, not a unitless fraction. The same goes for officials at the White House Office of Information and Regulatory Affairs who oversee federal cost-benefit analysis. Policy makers need dollars, lives, and jobs; 0.3 will not do.

In health-related fields, such as epidemiology, a common practice is to report estimated effects as an odds ratio. If an individual in the control group has a 0.0000010 percent chance of getting a certain disease, and individuals in the treated group (e.g., cell phone users, coffee drinkers, people who live under power lines) are calculated to have a 0.0000012 chance of getting the disease, the odds ratio is 1.2 (= 0.0000012 / 0.0000010). And so drinking coffee or whatever is being studied is said to increase the risk of disease by 20 percent. This seems large, but it is a misleading measure of magnitude by itself. Simply put, a 20 percent increase in an extremely small number is still an extremely small number. Creating categories of odds ratios that can be deemed small, medium, and large, as some have proposed,[12] has the same problem as all universal measures of effect size: in seeking to improve comparability, they standardize a number and remove it from the particulars of its context. Ironically, it is the particulars of the context on which questions of magnitude rest.

Guidance for Distinguishing Large from Small

Arguments about large and small cannot be reduced to calculating a t-statistic, an effect size, or any other statistic. Good arguments rest upon the careful combination of statistics, knowledge of particulars, and an awareness of values. They involve looking at a number from many angles and using the resulting breadth of perspective to form a line of reasoning that persuades us to view the number a certain way.

Assessing magnitude and arguing that something is large or small is not the same as arguing for a particular policy. To return to the map analogy from chapter 1, an argument about magnitude is like explaining the map to someone and the implications of what it shows. One may persuasively argue that the road to the mountains is long and arduous compared to the road to the beach without arguing that the beach is necessarily the best option. Similarly, one may persuasively argue that the achievement gap between Black and White students is disturbingly large without saying how much should be spent to close it or the best means to close it—whether it is greater school choice, more school funding, greater federal oversight of education, or something else.

Large and Small Problems

One form of assessing magnitude is to argue that something is large enough to be a problem and merit attention or action. Is it worth putting on the school board's agenda or the city council's agenda or the president's weekly policy-time agenda? In this sense, a large problem is one large enough to merit attention. But how large is too large? How much lead in the water? How many student suspensions from the high school?

When assessing whether a problem is large enough to warrant attention or action, consider the following:

- The audience and the magnitude of the problems competing for their time and resources
- The feasibility of doing something and the seriousness of doing nothing
- The assumptions embedded in your assessment

First, consider your audience and the magnitude of the other problems competing for their attention. Suppose that the number of students being suspended from high school per week increases from 1.0 in the previous semester to 1.5 students in the current semester. The school board does not have time to discuss all matters. The rise in suspensions must compete with other matters for the board's attention. There might be a million-dollar budget shortfall. The teachers' union might be threatening to strike. The magnitude of the rise in suspensions must be assessed relative to competing problems. If the teachers are threatening a strike, the rise in suspensions is a comparatively small concern. In quieter times, it might top the board's agenda.

To further illustrate the importance of knowing the magnitude of related and competing problems, imagine that you are researching sulfur in public drinking water, which makes water smell and taste bad and also causes diarrhea at high levels. You find that 2 percent of households have drinking water high in sulfur and want to argue that it merits the attention of public water authorities. It is hard to argue that 2 percent rate of high sulfur is a serious problem that warrants attention if lead is as common and has far more serious health effects. Even low exposure to lead can damage the nervous system of children and their ability to hear and learn. Simply put, the sulfur percentage is a relatively small problem compared to the lead percentage.

Whether something is large enough to merit attention is also audience-specific. For whom is the problem a problem? Whose attention should it command? The increase in suspensions might be too small for the board's attention but large enough for the assistant principal. Or to use the example of water quality, the geography of water-quality problems could be diverse and non-overlapping, meaning that some places have only one problem or the other. Public drinking water systems in older Midwestern cities may have a lead issue;

groundwater-dependent rural communities may have only a sulfur problem. Whether sulfur is problem, therefore, depends on context. Second, consider the feasibility of addressing the problem and the consequence of ignoring it. That 100 of 100,000 households have lead in their drinking water might seem ignorable. After all, it is *only* 0.1 percent of households. It has the look of small, but do not let appearances deceive you. No number is necessarily small or large based on how far it is from 0. What if the technology to remove lead is cheap and easy to use, and not removing it has a high chance of damaging the brains of exposed children? If so, 100 households drinking lead-laden water seems unacceptable given the feasibility of avoiding it.

Many studies report statistics that may seem important but are arguably ignorable. Consider a study published in *Science* that estimated that oil and gas infrastructure built from 2000 to 2012 in the United States and Canada occupies about 3 million hectares, roughly the equivalent of three Yellowstone parks. The study highlights that the occupation increases wildlife habitat fragmentation and heightens the need to conserve rangelands.[13] Aside from these amorphous implications, the consequences of ignoring land conversion are left vague. Will certain species likely go extinct? Will food prices rise because land is being used for well pads instead of growing wheat or grazing cattle? Or is this one more change, among a thousand others, that can be ignored with little harm to people, economies, or ecosystems as a whole?

Last, be aware of the assumptions embedded in your assessment that a problem is large enough to warrant attention or action. What is too large to ignore depends on one's view of the world. Those with greater optimism in the perfectibility of the individual and society will see as large what others might see as small and inevitable. Campaigns for zero suspensions, zero hunger, zero hate, zero greenhouse-gas emissions, and zero COVID will find nonzero numbers large enough to warrant a call to action. Any student being suspended might disturb some, while others see it as unavoidable.

Views on the role of government will also affect whether a problem is large enough for a particular audience to address. Suppose that 1 percent of households regularly skip meals because they lack money to buy food. One person might think that it is large enough to call to action civil society and individuals so they give their time and money to help their neighbors in need but not large enough to warrant action by lawmakers. Another person, with a different view of the role of government, might argue that it is large enough for lawmakers since the government's presumed role is to guarantee that basic needs are met.

Large and Small Effects

Some statistics quantify the real or potential effects of doing or having done something. They say that a 1-percentage-point increase in the tax rate on oil production reduces production by 5 percent. Consider the following four ways to assess the magnitude of an effect, all of which are complementary and together give a rich perspective on magnitude. Not all make sense in every situation, but they provide a mental checklist to work through as you assess whether an effect is large or small: (1) the effect in absolute terms, (2) the effect relative to the effort that caused it, (3) the effect relative to a baseline or a target, and (4) the effect's weightiness.

First, consider the effect in absolute terms. It is helpful to know the effect in absolute terms, both for each person or place affected and for the affected group as a whole. Absolute effects are effects in terms of people, dollars, tons, and the like. They say that better regulation of air pollution reduces by two the number of doctor visits per year for the average asthmatic child, reducing visits by 10,000 across the whole country. Absolute numbers do not provide all the perspective needed, but they are an important aspect of magnitude that people can easily grasp. They get what it means to avoid the cost and hassle of two doctor visits. Absolute numbers also feed into other measures of magnitude, such as the effect-to-effort ratio.

Second, consider the effect relative to the effort that caused it. In investment terms, what is the return per dollar or hour invested? More generically, what is the effect-to-effort ratio?[14] It would be absurd to conclude that $100,000 in profit is small without considering how much was spent to generate it—whether $200,000 in capital (a 50 percent return) or $20 million in capital (a 0.5 percent return).

Recall the fictional school-lunch example in which a small change in the timing of lunch increased math scores by 0.15 standard deviations. Because few people think in terms of standard deviations, suppose that the baseline math score equals the standard deviation, so the 0.15 effect size is a 15 percent increase relative to the baseline average math score. It is hard to say much about the magnitude of 15 percent apart from the magnitude of the effort behind it. In this case, the effort was a minor scheduling change that required no spending. In light of the tiny effort, a 15 percent improvement is an enormous effect.

Third, consider the effect relative to a baseline or a target. The effect relative to a baseline or target helps reveal the dent made in the overall problem or the progress toward the goal and complements the effect-to-effort ratio nicely. Suppose that a $1,000 per truck tax rebate spurs a doubling of hydrogen trucks in use (a large effect-to-effort ratio). As a result, the number of hydrogen trucks on the road increased from fifty to one hundred. This is a large increase relative to the baseline (100 percent!) because the baseline was so small. If the goal of the rebate was to put ten thousand hydrogen trucks on the road, we are only 1 percent of the way there.

Last, consider the weightiness of the effect. Here the focus is on the nature of the effect, how people's lives changed. Were lives saved? People kept out of prison? Consider the Moving to Opportunity program that gave select households housing vouchers so that they could move from high- to low-poverty neighborhoods. The effect of the move on youth participation in high school sports carries a different weight from its effect on participation in violent crime. It is a weighty thing to have people forgo committing violent crimes and stay out of prison. The weight of such life-changing outcomes is easy to grasp for their concreteness. Other outcomes, like increases in income, are harder to assess. Money is easy to measure, and most people would like more of it, but it is instrumental to other ends, some of which are weightier than others. Does the increase in income translate into a bigger television or better medicine for Grandma and a square meal for the kids?

Assessing the weightiness of one effect over another involves value judgments. Embrace offering such judgments as part of a transparent discussion of magnitude. The value of a thorough and explicit assessment of magnitude is to bring such judgments, which are often made implicitly, into the light of day. This brings them to people's attention so they can be thought about and discussed.

6

Think Hard about Causality

Conflating Correlation and Causation: A Recipe for Bad Policy

Most introductory statistics classes will teach students not to conflate correlation with causation. Ice cream sales are correlated with homicides but do not cause them.[1] Nor does a rise in umbrella sales cause it to rain. In addition, economics and other policy-related fields have gone through a credibility revolution, stirring enthusiasm for estimating causal effects and raising the standards for making causal claims from observational data.[2] Yet it is still worth underscoring why distinguishing correlation and causation matters for policies that affect whole organizations, cities, or groups. In policy, conflating correlation and causation is a good recipe for policies incapable of addressing the problem at hand and may make things worse.

A personal example will show how the conflation can waste time and money. During several heavy rainfalls one summer, a large pool of water accumulated above the drain behind my house and flooded my garage. This had never happened before. My neighbor attributed the overwhelmed drain to global warming and more intense storms, implying that the problem was the quantity of rain that had fallen and that was rushing to the drain. There was clearly a correlation—each storm had in fact flooded the drain. Though skeptical that the storms had been more intense than those in prior years, I adopted the basic logic that increased rainfall caused my flooding. I then shed sweat and money to replace half of the asphalt area surrounding the drain with grass that would allow the water to soak into the soil instead of rushing to the drain. The drain was nonetheless overwhelmed by the next storm. This suggested that the problem did not have its deepest root in the quantity of rainfall. With the drainage area reduced by half, the last storm would have needed to be dramatically more intense than any storm in prior years where flooding had not happened. This was not the case. The mystery of the overwhelmed drain was solved when a plumber explained the layout of my drain and showed me where to check for a clog. I did and found that muck had filled an important kink in the drain. I removed it, and the drain has since worked superbly.

In addition to promoting wasteful nonsolutions, conflating correlation and causation can lead to nonsolutions that make things worse, just like the wrong pill for a patient can make her more miserable. In my drain example, replacing asphalt with grass had the side benefit of creating more green space around my home. In other cases, applying the wrong solution could be harmful in addition to being wasteful. Numerous elite universities have stopped using standardized tests in admission decisions in hopes of expanding access to historically underrepresented groups. Admissions decisions will continue to be made but on other parts of the application. If overrepresented groups are better at finessing other parts of the application than they were at mastering standardized tests, the change could harm the groups the policy seeks to help. Time and good data and analysis will tell.

A True Causal Link Mismeasured by Correlation: A Pervasive Problem

It may seem obvious that correlation is different from causation and that confusing the two makes for bad policy. The most pervasive misjudgment related to correlation and causation is more subtle but just as problematic. The correlation-versus-causation examples commonly given in statistics classes obscure this pervasive problem with ice cream and homicide examples. Rarely does the policy maker do the equivalent of combating the rise in homicides by banning ice cream sales. It is more likely that the policy maker or her stakeholders are distracted by something causally related to the outcome of interest but not to the degree suggested by the correlation. Put simply, the correlation mismeasures the causal relationship, making us think that something is large when it is small or that it is small when it is large.

Consider the example of school funding and student achievement. Schools need to spend money to educate students, so there is an intuitive causal relationship between spending and achievement. Classrooms, heating and cooling, books, and teachers all cost money. In addition, spending per student is in fact correlated with various measures of student achievement, such as graduation rates and test scores.[3] But because school resources are highly correlated with other causes of student achievement, such as the home environment and family education and income, the raw correlation between spending per student and achievement overstates the actual effect. As a result, putting more money into the hands of administrators of low-resource schools will not close the achievement gap between the two groups of schools to the extent suggested by the simple correlation. Equalizing spending will not equalize achievement because students in previously low-spending schools will still not have the same home environment and resources as students in high-spending schools.

Conflating correlation with the magnitude of causation is more pervasive than you might think. It is common for a political leader to speak about the causes of a problem in a way that gives equal footing to small and large causes based on casual correlations. At the Council of Economic Advisers, I often reviewed factual claims in the draft speeches of White House officials. Speeches commonly included statements that were likely factually true but that conflated correlation with the magnitude of causation. The removal of several regulations was linked to a rise in oil and gas production, for example, two things correlated in time and causally related. Deregulatory actions happened before or were contemporaneous with a rise in production and almost certainly contributed something to it. But the simple correlation between deregulation and oil and gas production masks the magnitude of the causal effect of deregulation. Deregulation may have accounted for 5 percent of the increase in production, with the other 95 percent driven by rising oil prices. Casual and imprecise language presents the two causal factors as equally important or, worse, only mentions the policy change, which makes it appear to be the sole cause of higher production.

Conflating correlation with the magnitude of causation is not particular to presidential administrations or politicians in general. Anyone seeking to advance a policy wants to draw attention to particular causes. An environmental advocacy group highlights rising greenhouse-gas emissions and recent wildfires without discussing the effect of changing forest management. Or the steel-manufacturing lobby highlights a rise in industry employment following tariffs on imported steel, never mentioning expansion in the economy in general. Or a criminal justice organization highlights a reduction in bail amounts and a subsequent rise in crime, also without mentioning what else changed. In all the cases, a group uses a casual correlation to bring attention to one causal factor, treating all of the correlation as if it were causal when perhaps only a small fraction of it is. This can result in minor causes receiving the attention due to major causes, and to major causes receiving little or no attention.

Remember, this book is written not for the lobbyist but for the policy aide who gives the straight story to the policy maker she supports. This involves distinguishing major and minor causes. The aide can do much of this work with thoughtful use of the simple statistics from chapter 3.

Simple Statistics Can Often Uncover Inflated Causal Claims

Books on methods for estimating causal effects have abounded as part of the credibility revolution. As useful and important as the methods are, they are only part of the statistical analysis relevant to understanding causal effects in policy settings. Causal empirical studies generally seek to isolate and estimate the effect of one variable on another, X on Y apart from Z, school spending on student achievement apart from parental income. To do so, they look backward, using data from previous years, even many years ago, for which they can observe plausibly random changes in X and observe its effect on Y. Policy makers, in contrast, often look at current changes in Y and seek to understand what is causing the change. Candidate causes surface from all sides, and it is here that simple statistics play a powerful role by ruling out minor or irrelevant causes, perhaps narrowing the list of major causes to just one or two.

To uncover a minor cause masquerading as a major cause, recognize that for X to have driven a large change in Y, X itself must have changed by a sufficiently large amount. Police funding cannot explain much of a rise in crime if funding changed by only a percentage or two. Otherwise, it means that even a small change in police funding can have massive effects on crime, which is easy to rule out because funding probably rose or fell by several percentage points in prior years without creating sharp spikes or drops in crime. Thus, simple statistics such as the change in police spending can help to rule out bogus causal claims or prevent treating minor causes as major causes. Several examples will more fully illustrate the concept.

Returning to the example of my drain overwhelmed by rainfall, I could have avoided wasting my time and money on a nonsolution (replacing some asphalt with grass) if I had looked at a few simple statistics. I could have used local historical weather data to compare the rainfall of each of the recent storms with the most intense storm from prior years. Specifically, I could have compared the rainfall of recent storms with the maximum intensity of storms over the past five years. This would take data-organizing skills, but the statistics themselves are simple. Because my drain did not flood in previous years, the quantity of rainfall can explain my overwhelmed drain only if the recent storms had in fact dropped more rain than past storms. If not, then there is no increase in X, which is the quantity of rainfall, to explain the increase in Y, which is the quantity of water accumulating over my drain. If the quantity of rainfall had not increased, then something else must have changed, such as the capacity of my drain pipe, which was what had in fact changed.

The use of simple statistics in causal discussions can take many forms. The Opportunity Zone incentive created by the Tax Cut and Jobs Act gave investors a tax break for investing in selected low-income communities with the goal of reducing poverty and creating economic opportunity. It was signed into law at the end of 2017, and the selection of communities was complete by mid-2018. From 2017 to 2019, the number of people in the United States living in poverty fell by 5.7 million.[4] Now, suppose that at the end of 2019 an Opportunity Zone enthusiast had said: "The Opportunity Zone incentive is helping to lift Americans out of poverty. Since the incentive was signed into law, 5.7 million people have escaped poverty." Nothing is obviously factually wrong with the statement. Incentive-related investment probably lifted more than one person from poverty. It is also true that poverty fell by 5.7 million people over the period. Again, we have correlation and also a plausible causal link, but the correlation implies a far greater causal effect than what is actually the case.

Detective work and a few simple statistics can clear the fog created by the Opportunity Zone enthusiast, revealing that the incentive played a very small role in the reduction in poverty from 2017 to 2019. The communities selected as Opportunity Zones had a total poverty count of 7.5 million people to start with.[5] If the incentive caused a 10 percent decline in poverty in zones, which would be a sizable effect, it would account for only 13 percent of the total decline in poverty over the period (0.75 million / 5.7 million). Furthermore, the poverty reduction presumably happens as money is invested. Yet Opportunity Zone investment activity gained considerable momentum only in the second half of 2019.[6] And this is just money flowing into funds, not necessarily money flowing into contractor hands and creating jobs. Many projects can take a year or more of planning before shovels hit the ground. This means that the change in X (investment) thought to affect Y (poverty) was small. On the whole, then, a few simple statistics reveal that whatever poverty-reducing effect the Opportunity Zone incentive may have over its life, it played a very minor role in the fall in poverty from 2017 to 2019.

The Essence of Credible Causal Estimates

Methods for making a credible causal estimate can be sophisticated, and numerous books explain them clearly and with attention to their nuances, variations, and where using them is most appropriate. Methodologically, the simplest approach to estimating the effects of greater unemployment benefits, for example, is to randomly select in which areas unemployed people will have increased benefits and in which areas they won't. This is the essence of the randomized controlled trial, or RCT. By randomly assigning areas to the increased-benefit group and the no-increase group, the two groups should have similar characteristics on average. This includes characteristics such as the demographics of the population that might be correlated with higher or lower unemployment rates at present or in the near future. Because of the similarities on average, we would expect the two groups of areas to have similar changes in unemployment rates were it not for differences in unemployment benefits. If rates increased more in one group than the other, we would have reason to believe that the difference stems from the differences in benefits.

For many reasons, RCTs are not feasible for the policy aide. Regardless of whether our data reflect an actual experiment or what we observe naturally in the world around us (observational data), methods for estimating the causal effects of a policy generally have four main parts:

1. The change part: A change in policy, or whatever it is whose effect we want to estimate.
2. The cross-sectional part: Data on places (or people, firms, or schools) with a policy change and places with no change.
3. The temporal part: Data from before and after the policy change.
4. The counterfactual part: A reason to believe that the change for the policy-constant group shows what would have happened in the policy-change group had policy not changed.

It should be obvious that we need a change in the thing whose effect we want to study. But more is needed. Cross-sectional and temporal parts are needed, which involve data on places affected by the policy and those not affected (cross-sectional) and also data from before and after the policy change (temporal). To see why both parts are important, let's continue the case of unemployment benefits by considering the federal government's enhancement to unemployment benefits early in the COVID-19 pandemic. Believing that more generous unemployment benefits discouraged people from taking jobs, half of US states ended the benefit enhancement early, and half continued it until the federal government ended it nationwide.

Consider having only the cross-sectional part, the two groups of states at one moment. The moment could be from before half of states ended enhanced benefits or after it. If we have data only from before the end of the enhancement, we leave ourselves with no policy change to study, as all states at that time had enhanced benefits. If we have data only from after the change, we cannot separate the effects of the change from any preexisting differences in state unemployment rates. If states that cut the enhancement early, which I'll call cutting states, initially had unemployment rates higher than noncutting states, we might observe that cutting states still have higher unemployment rates after the cut even if the cut encouraged people to find jobs and reduced unemployment. We could always assume that the initial difference in unemployment was 0, but there is usually no good reason to assume it. (An exception is where the policy was purposely randomly assigned, in which case we expect similar average initial unemployment rates across the two groups of states.) And so we need a temporal component that permits differencing away any preexisting differences in unemployment rates across the two groups of states.

Or consider having only the temporal part, which would be the case if the entire country cut the enhancement at the same time. This is also problematic. We could look at the unemployment rate before and after the cut, but this would not provide a credible estimate because so much else in the economy was changing at the same time—travel restrictions, global supply changes, and more. Just as the cross-sectional-only comparison required the assumption that both groups had the same average unemployment rate before the cut, the temporal-only comparison requires the assumption that there would have been no change in the unemployment rate had the cut never happened. This is unlikely.

To illustrate how the cross-sectional and temporal parts work together, consider the following mean unemployment rates: the rate among cutting states before the cut ($U^{B,C}$) and after the cut ($U^{A,C}$), and the rate among noncutting states before ($U^{B,NC}$) and after the cut ($U^{A,NC}$). Table 6.1 shows the four means and the two cross-sectional differences and the two temporal differences. The differences by themselves have the limitations just discussed, but combining them is powerful. Table 6.1 shows this combination of the cross-sectional

Table 6.1 Cross-Sectional and Temporal Differences and Their Combination

States/timing	Before cut	After cut	Horizontal difference	Description
Cutting states	$U^{B,C}$	$U^{A,C}$	$U^{A,C} - U^{B,C}$	Temporal only (with a policy change)
Noncutting states	$U^{B,NC}$	$U^{A,NC}$	$U^{A,NC} - U^{B,NC}$	Temporal only (without a policy change)
Vertical difference	$U^{B,C} - U^{B,NC}$	$U^{A,C} - U^{A,NC}$	$(U^{A,C} - U^{A,NC}) - (U^{B,C} - U^{B,NC})$	Difference-in-differences
Description	Cross-sectional only (pre-policy differences)	Cross-sectional only (post-policy differences)	Difference-in-differences	

and temporal differences into the difference-in-differences. It takes the post-cut difference across the two groups of states ($U^{A,C} - U^{A,NC}$) and subtracts from it the pre-cut difference ($U^{B,C} - U^{B,NC}$), which was the major limitation of the cross-sectional-only comparison. Alternatively—and equivalently—the difference-in-differences could be rearranged as the temporal difference for the cutting states ($U^{A,C} - U^{B,C}$) less the temporal difference for noncutting states ($U^{A,NC} - U^{B,NC}$). This takes the temporal difference for the cutting states and removes from it any change that would have occurred over time independent of the policy change, which was the major limitation of the temporal-only comparison.

Thus, the two differences, or what is often called the difference-in-differences, remove two main sources of bias in causal estimates: that places are different from each other and that much can change in a short time. Accounting for places being different at one moment (cross-sectional differences) and for commonly experienced changes over time (temporal differences) goes a long way toward making a credible causal estimate.

Yet one part remains—the counterfactual part. This involves considering the likelihood that the two groups would have had the same change in unemployment rates had they adopted the same policy. This is important because when we compare the change in the unemployment rate across the two groups and treat the difference as the causal effect of the cut, we are in fact assuming that the two groups would have had the same change in unemployment rates had they had the same policy. Put differently, noncutting states are used to peep into an alternative universe to see what the unemployment rate would have been in cutting states had they not cut the enhancement. If the federal government had randomly selected some states in which to cut benefits early, we would have reason to expect the two groups to have the same evolution in unemployment rates in an alternative universe where both keep the enhancement. The situation is less clear when states themselves select what group to join. Maybe states expecting a greater rebound in their economy chose to cut benefits and those expecting a smaller rebound chose to keep them as long as possible.

The main way to explore the credibility of the noncutting states as the counterfactual is to look at the pre-policy-change trend across the two groups. Did cutting and noncutting states experience a similar change in unemployment rates in the months before the cut? If the changes were different despite having the same policy during the time, noncutting states are probably not a good counterfactual for cutting states. After all, they were clearly not a good counterfactual before the policy change! Note that looking at prior trends is important even in an RCT because randomization works on average, but any two randomly selected groups might be quite different by chance.

Even if the pre-policy-change trends were similar, post-policy changes in the economy might have affected the two groups differently, which would also undermine the use of noncutting states as the counterfactual. But forces at work over time often affect similar places in the same way. Two agricultural areas are more likely to experience similar economic trends than are an agricultural area and an industrial area—one being affected by a rise in the price of corn and the other not so much. In the unemployment example, this would take the form of looking at the characteristics of cutting and noncutting states. Did they previously have similar levels of unemployment? Were they similarly distributed across the country, or were all cutting states in the South and all noncutting states in the North? And so on. Here it is helpful to think about what caused the two groups to pursue different policies in the first place and whether the cause might be related to unemployment trends apart from any change in benefits.

Journalistic Difference-in-Differences: Helpful Causal Estimates Are Often Nearby

Academic studies estimating causal effects with observational data often include a battery of statistical tests and rabbit trails meant to bolster the credibility of the estimates. But the core parts of credible causal estimation (e.g., the change part, the cross-sectional part) are easy to grasp and apply, so much so that they often appear in newspaper articles for broad audiences. I call these simple, accessible, and intuitive applications "journalistic difference-in-differences." The assessment of prior trends and its main takeaway can be told with a simple figure. This does not require sophisticated statistical skills to do and even fewer skills to understand. And it is what a policy aide often needs—quick-to-make and easy-to-digest estimates that give a rough sense of the magnitude of the true effect.

Journalists at the *Wall Street Journal* provide a nice example of journalistic difference-in-differences. In an article on the effects of enhanced unemployment benefits, they look at unemployment rates before and after half of US states had announced or actually cut the enhancement.[7] The core of the *Journal*'s analysis is a figure showing the unemployment rate over time, before, during, and after the cuts, with separate lines for cutting and noncutting states. The advantage of the graphical display of unemployment rates over time is that it permits a visual assessment of prior trends in unemployment rates across the two groups. Although most readers probably do not use the language of counterfactuals, they get that if the unemployment rate lines for the two groups are diverging before the cuts, then there is good reason to expect the lines to continue to diverge regardless of any actual effect of the cuts. For January 2021 until July 2021, figure 6.1 shows the same data and state classification used for the *Wall Street Journal* article.

The figure shows that cutting states had an overall lower unemployment rate throughout the period and that both groups of states

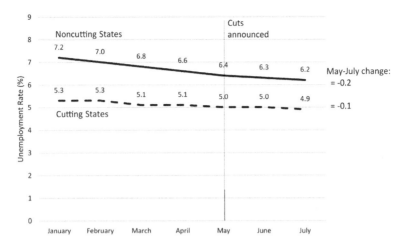

Figure 6.1 Unemployment rates for cutting and noncutting states. The unemployment rates and the classification of states are the same as those used by Cambon and Dougherty in "States That Cut Unemployment Benefits Saw Limited Impact on Job Growth," *Wall Street Journal*, September 1, 2021.

experienced falling unemployment rates. Policies regarding enhanced benefits began diverging in May, so the relevant difference is from May to July, which is the last month of available data at the time of the article. The main takeaway from the figure is that both groups of states experienced roughly similar changes in the unemployment rate from May to July. Looking at the actual numbers shows that cutting states actually had a smaller drop in the unemployment rate than noncutting states, –0.1 compared to –0.2. The difference-in-differences estimate therefore suggests an increase of 0.1 percentage points in the unemployment rate for cutting states relative to noncutting states (= –0.1 – [–0.2]).

The trends prior to the policy change, however, reveal a larger April-to-May drop in the unemployment rate for noncutting states. If the trend had continued in June and July, noncutting states would have had a 6.0 percent unemployment rate by July (6.4 in May, 6.2 in June, and 6.0 in July), whereas cutting states would have had a 4.8 percent rate (5.0 in May, 4.9 in June, and 4.8 in July). So, cutting states had an unemployment rate 0.1 percentage points above their expected level (4.9 instead of 4.8) compared to the noncutting states that were 0.2 percentage points above their expected level (6.2 instead of 6.0). Accounting for prior trends, then, suggests that the benefit cut did not increase the unemployment rate and may have decreased it very slightly.

Despite being for a national newspaper with severe space constraints, the *Journal*'s analysis includes a type of robustness analysis that explores whether a similar result emerges when using a related but distinct outcome. The unemployment rate is a ratio, with the numerator being the number of people unemployed but searching for work and the denominator being the number of people in the labor force (who may or may not be working). Total nonfarm payrolls is just a count of the number of jobs in the economy outside of the farm sector. A second figure in the article therefore shows the percentage change in payrolls from May to July for cutting states and, separately, for noncutting states. Payroll growth tells the same story as the change in the unemployment rate, with cutting and noncutting states having nearly identical payroll growth (1.33 percent compared to 1.37 percent).

Is Journalistic Difference-in-Differences Rigorous Enough?

By the time academic studies are public, let alone peer reviewed and published, the window for a policy decision has often passed (although it may reopen in five years). For example, the earliest academic study of the effect of the cuts on unemployment was released in December 2021, three months after the nationwide termination of enhanced benefits.[8] More rigorous analysis might yield different estimates, but probably not considerably so. In the case of enhanced benefits, researchers later estimated that the national unemployment rate would have been 0.3 percentage points lower in July and August 2021 had the entire country cut the enhancement.[9] Differences between a quick journalistic difference-in-difference and an academic difference-in-difference will likely be small compared to the range of effects speculated at the time of the policy decision. When someone is speculating about a 2-percentage-point effect, the difference between 0.1 and 0.3 is like the difference between beige and khaki (for all but those with the most sensitive sense of color).

Although journalistic difference-in-differences may not pass the bar of top academic journals, it can greatly improve upon what is known at the moment. Academic articles cover only certain topics, and articles often appear too late to affect policy decisions. It makes little sense to blindly apply academic standards as if they magically weighed timeliness, relevance, and credibility the same as a policy maker or the people he represents. For example, it matters which claims people are making and how much analysis and evidence is available.

Imagine that the enhanced benefits were to last a year longer absent action by the states and that some states had moved quickly to cut them. Consider a governor of a state that had not yet cut them but who would like to see the unemployment rate return to its prepandemic level sooner than later. Suppose that some people with the governor's ear are speculating that enhanced benefits are the main reason the unemployment rate is still so high, suggesting that the rate could fall by 2 or 3 percentage points if benefits were cut. This might be pure speculation or perhaps speculation mixed with findings from an academic study of a policy experiment in Japan in the aftermath of the 2008 financial crisis. In this case, a journalistic difference-in-differences (JDID) estimate of, for example, 0.3 percentage points adds clarity to the conversation like a floodlight to a candlelit room. It matters little if the true effect is 100 percent larger (0.6 percentage points) because both help to cast doubt on the claims about a 2- or 3-percentage-point effect.

Moreover, the governor and everyone in the room will understand the JDID estimate, and the speculators suggesting a 2-percentage-point drop will have to deal with it. The JDID estimate is often all that is needed to reach the right big-picture conclusion, as it was in the case of enhanced unemployment benefits. As the University of Chicago economist Peter Ganong put it, "If the question is, 'Is [unemployment insurance] the key thing that's holding back the labor market recovery?' the answer is no, definitely not, based on the available data."[10]

The standard statistical significance protocol has the academic fretting over whether enhanced benefits had a zero effect or something like a 0.3-percentage-point effect. The governor, in contrast, probably treats both scenarios as leading to the same decision (keep the benefits). The governor wants to know if the effect is closer to 0 than it is to 2 or 3 percentage points. In the language of the loss function, the academic is considering small prediction errors (0 versus a 0.3-percentage-point decrease), and the governor, large prediction errors (roughly 0 versus 2 to 3 percentage points). More importantly, the governor is also trying to translate prediction errors into a policy error and the various real-world costs such a policy error could have.

Note that the standard error and 95 percent confidence interval for the estimated difference in unemployment rates measures only uncertainty stemming from sampling error, that is, error stemming from data not representative of the population. It does not speak to error stemming from a bad research design that poorly accounts for potential confounding factors.[11]

One critique of analyses like JDID is that they use observational data, not data from a randomized controlled trial (RCT). In policy, as with journalism, timeliness is often paramount, so one works with what is at hand. A strand of the credibility revolution reserves the highest marks of credibility and importance for estimates from RCTs. And just as some insist upon having a random sample to calculate a useful statistic, some insist that only an RCT can be relied upon to produce truly credible estimates of causal effects.

Yet as I argued in chapter 2, a dogmatic focus on random samples has misguided many, and one can give too much deference to RCTs. As the Nobel laureate Angus Deaton put it: "Just as none of the strengths of RCTs are possessed by RCTs alone, none of their weaknesses are theirs alone. . . . There is no gold standard. There are good studies and bad studies, and that is all." In the same article he endorses Martin Ravallion's conclusion about methods, which Deaton summarizes as "the best method is always the one that yields the most convincing and relevant answers in the context at hand."[12]

Answers that are convincing and relevant in the context at hand have two parts. Both parts are important. A convincing answer that is irrelevant in the context at hand is just that—irrelevant and therefore ignorable. A way to be irrelevant is to be untimely. This is often the weakness of academic work and a strength of the policy aide's JDID analysis. And sometimes not even data have accumulated for even a JDID, in which case a theory and model will be needed to project likely outcomes (more on projections in chapter 8).

An answer that is relevant but not convincing may also be ignored depending on the context. Policy makers, who are in a position of trust and authority, do not want to base decisions on unconvincing evidence that is so lacking credibility as to be laughable. Both credibility and relevance matter, but the appropriate weight to give to each is itself context specific. Academic standards, especially in disciplines such as economics, give more weight to credibility than to relevance. This leads to looking for answers only in places where light shines brightly, even though many relevant answers lie elsewhere. But if you find an important answer in a dark place, meaning an answer that is only mildly rigorous and credible, how much should you discount it? As usual, it depends on context.

Irreversibility, Credibility, and Constancy

The value of less biased or more precise estimates varies greatly. In some cases experimenting with a change in policy brings little risk, and all the evidence needed is that which shows that success is more likely than not. A low bar of evidence can make sense if the old policy can be restored without creating confusion, damaging credibility, or other harms that persist beyond the policy reversal. This may be the case for small-scale changes that affect only one department or one locality. The more-likely-than-not standard contrasts with academic standards for research design, which can readily discard any estimate with some bias from a confounding factor.

In other settings, a good research design might not be enough. We might want to distinguish between 0 and something very close to 0 but not 0. Common statistical practice, with or without an RCT, focuses on whether we can rule out 0, not on whether we can rule out something very close to 0 but not 0. Consider the difference between a drug with a 0 percent chance of making you go blind and one with a 0.1 percent chance of doing so. Or think of roller coasters and the probability of failure or the presence of lead in drinking water. These differences matter. Some effects are irreversible, so we want great confidence in the safety of our drugs, our drinking water, and our roller coasters. The more irreversible the damage from things going wrong, the more confidence we want in our decisions regarding the potential negative effects of taking the drug, drinking the water, or boarding the roller coaster.

In still other settings, we might want to be really sure that a policy has a positive effect, and so the focus on research design and statistical significance can make sense. The credibility of people and institutions built over years can be lost in a day. Telling people that something works and then saying that it doesn't erodes credibility. People also get fed up with needless policy flip-flopping. Residents, builders, and everyone else like constancy. Residents who have always recycled glass, plastic, and paper in blue bins on Tuesdays will be thrown off by a new practice of recycling only paper and plastic and doing so in yellow bins on Fridays. The same can be said for home builders familiar with a particular timeline and process for site permitting and building inspections. Change can create confusion and inconvenience. More importantly, people and businesses regularly make decisions with long-term implications—where to live, whether to buy a house, whether to invest in an expensive piece of equipment. The decisions are reversible but at a cost: the cost of moving houses again or selling a piece of (now) used equipment.

Statistics Are Helpful but Rarely Decisive

The credibility revolution and its emphasis on more credible causal estimates can give the impression that developing a bullet-proof estimate is 95 percent of what is needed for good policy. With uncertainty over causality expunged from the estimate, policy implications will flow effortlessly and obviously, as if one had already written the recipe, bought the ingredients, and mixed them together in the pot so that others need only turn on the stove and set the timer. But as chapter 1 explained, statistics are rarely decisive for policy even if there is little uncertainty about them.

Decisions require the weighing of trade-offs, and the weights rarely come from the data. In the unemployment example, the main trade-off was between reducing the unemployment rate and forgoing payments from the federal government to state residents already unemployed, money that will simply remain in the Treasury if the state does not take advantage of it. These are not the only outcomes involved, because changes in both will have their own downstream effects, such as implications for business hiring and tax revenues. And of course the decision has political dimensions as well. The party that approved the payments at the federal level might be different from the governor's party, so taking the money can be seen as endorsing the policy choices of the opposing party.

The economist Jean Drèze makes a similar point with more color. He considers a randomized controlled trial where researchers experiment with adding eggs to school meals in India so as to improve student nutrition and learning:

> No value judgments are required to conduct an RCT aimed at examining whether adding eggs in school meals helps to enhance pupil attendance or child nutrition. But advocating the inclusion of eggs in school meals is a very different ballgame. It means dealing with the arguments of upper-caste vegetarian lobbies (eggs are considered nonvegetarian in India) and animal-rights activists, aside from those of the Finance Ministry, the Education Department, and teachers' unions. Commercial interests, too, are likely to come into play as the poultry business eyes big contracts. The debate can easily get very charged. Any "advice" offered in this charged atmosphere may have serious repercussions, good or bad. A worst-case but not uncommon scenario is that a piece of advice turns out to be counterproductive. Dealing with these choices, conflicts, and dilemmas requires much more than "evidence."[13]

The word *evidence* could be replaced with the phrase "a credible estimate of causal effects." Drèze's point is similar to the one made in chapter 1, that the road from data to decisions is often long and tortuous. Everyone might agree that the data show that students who received an egg a day as part of the study had better attendance than those who received no egg. But start recommending that this be done on a large scale by the Ministry of Education using taxpayer funds, and the conversation will quickly assume a new liveliness. Remember, statistics is like mapmaking, not destination setting. To use the map to select a destination is to enter a different, messier realm.

7

Show That You've Been to Table School

In a hastily called White House meeting, several handouts with tables of statistics were distributed to the ten or so people in the room. A research assistant had created them on short notice, and in his haste he had left the statistics in dollars. Because the values were in the billions, this meant that the tables were full of twelve-digit numbers like $121,657,782,910. (We were spared decimal places, if I recall correctly.) Within moments of receiving the handouts, the senior official in the room loudly remarked that whoever made the tables needed to go to table school. The official was so distracted by the multitude of numbers that he commented on their absurdity several times throughout the meeting. The tables' author was in the room, and most people present knew it but in mercy said nothing, acting as if the tables had fallen from the sky.

For you and your statistics to be taken seriously, you must show that you've been to table school. The core curriculum of table school covers principles for the clear and undistracted communication of statistics. The point of table school is that your statistics and tables and figures can be understood with little effort and cannot be misunderstood.

Deirdre McCloskey, Distinguished Professor Emerita of Economics, History, English, and Communication, approvingly quotes a famous Roman as saying that one ought to write so that the reader cannot possibly misunderstand.[1] The same applies to the presentation of statistics. During my early months at the Council of Economic Advisers, I created a figure that the White House subsequently tweeted. In a short time, the council chairman heard from people confused by the figure. My first response was to blame them for not understanding such a clear figure, but the chairman agreed with the confused. I had to do better. I then spent most of a day reworking the figure, rethinking the labels and every detail that could aid or thwart comprehension of the figure. To follow McCloskey, I was tasked not with creating a figure that the chairman and I could understand—the original figure met that bar. My task was not even to create a figure that a broad audience could understand. The task was to create a figure that our audience could not possibly misunderstand.

The Value of Table School

Master the principles of table school so that your statistics create understanding in the minds of your audience. Your statistical savvy will not create understanding if it is not matched by the ability to communicate. One theme of this book is that calculating the right numbers is only half the job. The other half involves placing the right adjectives next to the numbers and having reasons for them (see chapter 6). It also involves presenting them clearly, either in text or in tables or figures. This can require as much time and effort as generating the numbers themselves and is equally important to do well.

For your statistics to further understanding, the audience must pay attention to them, understand them, and believe them. The first challenge is to capture the audience's attention and keep it long enough to enable understanding. Meeting the challenge requires clear and simple figures or tables. Academic audiences value clarity in presentation, but they also have a high tolerance for complicated and poorly labeled tables and figures. Recognizing the difficulty of simplifying, people who have created bad figures, which includes almost every academic, often empathize with others when they do the same. Policy audiences don't have the same tolerance or empathy. They are too busy, and the competition for their attention is too intense. If clear and simple figures and tables often fail to attract and retain eyeballs, busy and opaque figures don't have a chance.

The audience is more likely to attend to your figure if it looks intelligible. But we want to achieve more. We want them to actually understand it and not possibly misunderstand it. After all, a misunderstood right number can mislead as much as a wrong number. The audience must grasp the units, the population considered, and the overall point—that average debt per graduating student has increased much more among graduate students than undergraduate students, or that the number of emergency room visits related to asthma has fallen less in areas with coal power plants than in areas without them.

Last, for a statistical point to affect understanding, it must be believed. Points understood and disbelieved do not affect understanding. The mind grasps them and then releases them. For people to believe your points, you must convey credibility, much of which comes through tables and figures. They convey the thoughtfulness, diligence, and trustworthiness of their creators. You must show competency in visible things to earn trust regarding invisible things like behind-the-scenes data manipulations and calculations. After all, if by mistake a word is omitted from a figure title, perhaps the figure's author also missed a step in cleaning the data or calculating the statistic. And if the figure has an awkward label or two more decimal places than needed, perhaps the author gave little thought to the conceptual soundness of calculations and their interpretation.

Table School Principles

Know Your Audience and Purpose

The first principle of table school is that knowledge of your audience and purpose should guide the application of all other principles. Figure 7.1 depicts the spectrum of audiences and purposes. On one end is an audience of fellow policy wonks, and at the other end is a broad audience, which may include one person with broad responsibilities (the mayor) or an actually broad audience like the media and the public. Your purpose for engaging with each audience varies, for example, to deliberate versus to inform. How you engage also varies, for example, providing more versus less detail.

Starting at the left end of the spectrum, policy wonks are people in your organization or peers in other organizations who are steeped

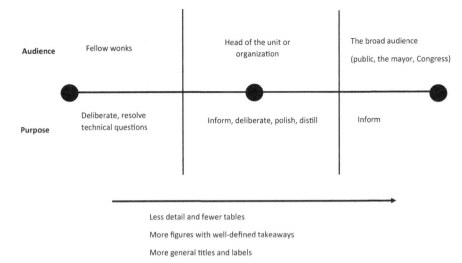

Figure 7.1 The audience-purpose spectrum

in the methods and subject. An example of the wonk level is a meeting with specialists in a division of the state's Department of Health about the workings of one of the department's programs. The specialists know the workings of their program better than you and all the caveats of the data because they produce it or work with it daily. The purpose of sharing information with them is to spur conversation and deliberation in order to better understand the issues. Here, more breakouts of the data—by group, by year, by outcome—are valuable. More technical titles and labels are appropriate because your audience knows ten ways to measure a person's stress level, a student's proficiency in math, or an economy's inflation rate. The figures have more lines and bars and less clear takeaways since a purpose of engaging with the group is to help identify the takeaways.

The next node on the spectrum is where the audience might be the head of the Department of Health and her chief of staff. They are familiar with the program under study but not like the specialist policy wonks. The program might be one of ten that the department oversees. The wonks would happily spend two hours talking about data or the program's effectiveness, but the department head is available for only a half hour. Although she might have been a policy wonk immersed in the details at one time, her higher-level position requires that she think about more issues in less time. Your communication, therefore, must be more polished and distilled than when engaging with the wonks. It should use fewer tables and more figures that are simple and pleasing to the eye. The purpose of meeting with the head is to educate her and to deliberate with her about what other information is needed and which cuts and refinements to make, particularly if the head is to then present to another, higher-level audience of policy makers or to the media, to brief the mayor or governor, or to testify in Congress.

The last node is when communicating with the broadest and probably most important audience—the media, the mayor or governor, or Congress. This might be you or the agency head presenting about the program to a committee of the State House of Representatives. The presentation addresses people much less familiar with the program, with attention scattered across a myriad of topics, and who are perhaps even busier than the agency head. You might have only five or ten minutes to convey your points. At this level the purpose is to present and inform so that the audience, not you, discovers. Every figure needs a clear takeaway point and must include only the elements necessary to understand the figure and the takeaway. Titles and labels should favor readability over precision. Everything must be polished to ensure ease of comprehension for a busy audience.

My experience at the Council of Economic Advisers spanned the spectrum shown in figure 7.1. Team members worked independently or in pairs to collect and sift information. We then met as a team, perhaps for an hour, to further wrap our minds around the topic (the wonk audience). Sometimes the teams included knowledgeable peers from other entities in the building, such as the Office of Management and Budget or the Domestic Policy Council. The team then produced a memo or handouts for a meeting with the council chairman (the unit head audience). The chairman did not have an hour to ponder ten tables. He usually wanted the takeaway points and associated figures upfront and then asked questions to convince himself that these were indeed the key takeaways and figures. The outcome of the meeting might be instructions for preparing a slide deck for a peer (other cabinet members or senior advisers to the president) or the president (the broad audience). Then the slide deck would typically pass through several iterations internally and then, if going to the president, further review by the president's staff secretary.

Where you find yourself on the audience-purpose spectrum should guide the application of the three principles of table school that follow. The principles apply to all situations but to different degrees depending on the audience and purpose. For example, "Present Only What Is Needed" looks differently if your audience has five minutes or an hour.

Present Only What Is Needed

In a good sentence every word has a job and cannot be cut without loss of clarity or substance. The same principle applies to the communication of statistics, whether it be statistics in a paragraph, the bars or lines in a figure, the columns and rows in a table, or even the digits in the statistic itself. Every statistic, line, and bar requires effort to understand. If the task appears too demanding, the audience will gloss over the material. Ask for too much attention, and you may receive none.

Imagine reading that 50 percent of young children in Somalia were malnourished in 2022. Understanding the statistic involves grasping that it is for Somalia in 2022, and it is a percent, with the denominator being young children and the numerator being children who are malnourished. More effort goes into absorbing the reality reflected by the statistic that one in two children—children like your son or nephew or granddaughter—have too little food to be healthy. That is a lot to absorb. Thus, a paragraph with three statistics is asking much of the audience; one with six almost guarantees that the audience will gloss over them.

Do not lessen the focus on an important statistic by surrounding it with tangential statistics. Continuing with the paragraph example, if the paragraph's point is that there is a hunger crisis in Somalia, then the percentage of children malnourished makes the point, perhaps complemented by the increase from a prior year or a comparison with the regional average. We do not need statistics on poverty or crop production or food imports. They might be relevant somewhere in the memo or presentation, but they should not be dumped on the audience merely because you discovered them and they struck you as

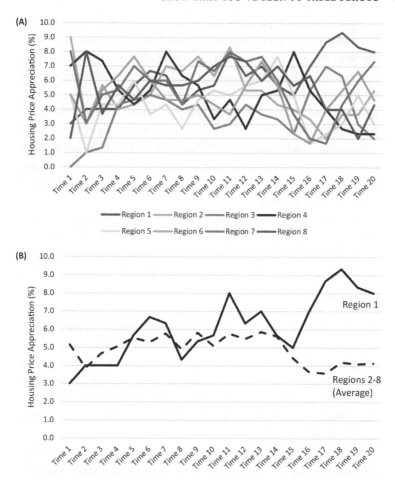

Figure 7.2 With figures, less is usually more. (A) An unpleasing figure. (B) A figure in which the point pops.

generally relevant. If the point is that there is a hunger crisis, pick the best statistic to illustrate it and clear away any distractions around it.

Just as every statistic in a paragraph should be indispensable, so too should every line or bar in your figure and rows and columns in your table. If your point is that home prices in one region of the country have grown much more over the past five years than everywhere else, spare the audience the tedium of following eight lines over the years, one for every region (fig. 7.2a). Show the line for the region of interest and one line for the other seven regions (fig. 7.2b).

Apply the same principle to bar charts—keeping only the bars necessary for the point, where different bars often represent different periods (2010-2014 or 2015-2019) or different groups (eastern and western halves of the country). And do the same for tables. Rows often represent different variables (life expectancy, income, wealth), and columns represent different statistics (mean and median) or different periods or groups. Convince yourself that every element of the table matters and cannot be simplified without losing the point. If you can't convince yourself, cut and simplify. Report only the median, drop income and keep wealth, and use only two periods instead of three. You get the idea.

The principle of presenting only what is needed applies to the number of digits reported for a statistic. Returning to the story of the handout with twelve-digit numbers, what should have the research assistant reported? The question concerns the proper scale for reporting. Milligrams or kilograms? Thousands or billions? Writing 121,657,782,910 is distracting and nearly indigestible. But what about 121,657,783 thousand or 121,658 million or 122 billion? If the year-to-year variation in the statistic is several billion dollars or more, anything beyond billions is distracting, so report 122 billion instead of 121.7 billion. Imagine that instead of the table having numbers like 121,457,782,910, the assistant had instead reported 12.1457782910 billion. The response by the official would have been the same: "Off to table school!"

Those who have not attended table school often report all the decimal places given by the calculator or statistical software. A similar mistake made by those who have forgotten the principles of table school is reporting as many decimals as is conventional. Both mistakes usually force the audience to look at more numbers than necessary, which violates the principle of presenting only what is needed. The unnecessary numbers distract the audience and suggest that the author behind the numbers is a button pusher, not a thinker.

Yet choosing the right number of decimal places to report—whether in text, a table, or a figure—is less straightforward than you might think. It requires understanding units (chapter 2) and assessing magnitude (chapter 5). Consider data on lead concentrations in drinking water in Pittsburgh. In 2016, the city's water authority was required to sample drinking water for lead. It sampled one hundred homes and shared the results in a letter to residents. On the basis of the results reported in the letter, which I received as a Pittsburgh resident, I calculated a mean lead concentration of 8.63 parts per billion (ppb).[2]

In a memo to the mayor, should you report 8.63 or 8.6 or 9? I recommend 9 for several reasons. First, variation in the data is large relative to the 0.4 units added in rounding. The standard deviation, for example, is 14 ppb. Second, 0.4 is small relative to the precision of the measurements. The letter with the data classified all readings less than 2.1 ppb as "nondetect," suggesting a fair degree of imprecision measuring lead. Last, 8.63 and 9 are roughly the same distance from the most policy-relevant number, 15 ppb, which is the Environmental Protection Agency's (EPA) action level for lead.[3]

The existence of a lead action level suggests that another number is more relevant than the mean concentration: the percentage of homes with a lead level exceeding the action level. It turns out that 17 percent of sampled homes had a concentration exceeding the action level. Had the city not sampled exactly one hundred homes, the percentage might have been 17.32 percent. Again, telling the mayor that 17 percent of sampled homes had lead above the EPA action level is better than telling him about the "seventeen point three two" percent of households (note the three extra words needed to say 17.32). Knowledge of the subject matter is also helpful here. The EPA requires water authorities to take action if more than 10 percent of homes sampled exceed the action level. Clearly, Pittsburgh's water authority must do something if its percentage is 16.6 percent or 17.4 percent. Give everyone's eyeballs and tongues a rest, and report 17 percent.

Of course, additional decimal places have their use, too. An increase of one-tenth of a percentage point can translate into billions of dollars or thousands of lives. In the Olympics, several hundredths of a second can separate the gold medal winner from the no-medal winners. The key is to grasp the practical implication of an additional tenth or hundredth, and report as many digits as needed and not one more.

To exercise your discernment in reporting, consider another example. Median household income in the city increased by $1,107 from the prior year. In a presentation to the mayor, should you write $1,107 and say "one thousand one hundred and seven dollars"?

Probably not. The mayor and the press will more easily absorb and remember seeing $1,100 and hearing "eleven hundred." The mayor won't receive Pinocchios from the fact-checkers when he talks about the eleven-hundred-dollar increase. He is not a statistical agency producing a statistical report. He is a leader intelligently communicating to normal people about the scale of economic change occurring in their midst.

Use Readable and Adequately Precise Titles and Labels

Figures and tables need titles and labels for their axes, lines, columns, and rows, all of which should further understanding and prevent misunderstanding. Titles should be readable and provide an adequately precise description of the content of the table or figure. This requires more effort and thought than you might think. Consider four different titles for a figure of housing price appreciation by region for the years 2010–2019:

1. The annual appreciation rate in the FHFA estimated median housing price for one-unit noncondominium properties by region, 2010–2019
2. The annual appreciation rate in the FHFA housing price index by region, 2010–2019
3. Annual housing price appreciation by region, 2010–2019
4. In the 2010s, home prices increased more in regions that most regulate land use

The first title precisely conveys what the appreciation rate is based upon: the FHFA estimated median price for one-unit, noncondominium properties. But it sacrifices readability for precision. The second title is more concise but retains the distracting acronym (FHFA) and the idea of an index, which can raise distracting questions: What is an index? How is it constructed? The third option retains some precision and gains readability.[4] Note that it drops the term *rate*, which arguably improves conciseness without losing clarity. *Appreciation* can mean the dollar-value or percentage change in the value of something. It will be clear that appreciation is presented as a rate if the vertical axis has the label "Percent."

Depending on your audience, you may want the title to state

the figure's takeaway message. The fourth title does this, assuming that the lines for the regions are labeled as high- and low-regulation regions. The more receptive the audience, the more they will appreciate a punchy and conclusive title. The more skeptical the audience, the better a purely descriptive title. Help the skeptics reach the takeaway themselves without feeling coerced.

Labels must also be thought through like titles and cohere with them. If the title uses "annual housing price appreciation," the vertical axis should follow suit and repeat it or a portion of it (e.g., "Housing price appreciation, %"). More concise is better as long as the label adequately describes the variable or group. Here again there is a trade-off between precision and readability. To grow in discerning a good balance between precision and readability, study the figures of newspapers like the *New York Times* and the *Wall Street Journal*. Their broad audience includes policy makers. As newspapers of record, they strive for accuracy, but being readable to a broad audience is also a must. This includes the readability of titles and labels.

Use Pleasing, Simple, and Faithful Figures

An essential tool in communicating statistics is a figure that pleases the eye and is simple and faithful. The format (figure) and its attributes (pleasing to the eye, simple, faithful) help the audience understand and never misunderstand statistical takeaways.

Figures tell stories more clearly than tables do. Two lines moving together over time and then diverging in the year of a policy change affecting one and not the other speaks more clearly than a table with two columns of numbers. The *New York Times* and the *Wall Street Journal* contain many figures and few tables. The same is true for briefing materials for policy makers, and the higher up the position of the policy maker, the less likely that she sees a table. I cannot recall one table among the slides I reviewed or prepared for the president or cabinet-level officials.

To leverage the natural advantage of figures, they must be made to please the eye, which makes the figure's takeaway point pop. Consider the annual average temperature at the Honolulu airport in Hawaii over time. The annual variation yields a line that looks like shark teeth (fig. 7.3a), making it hard to see that temperatures have

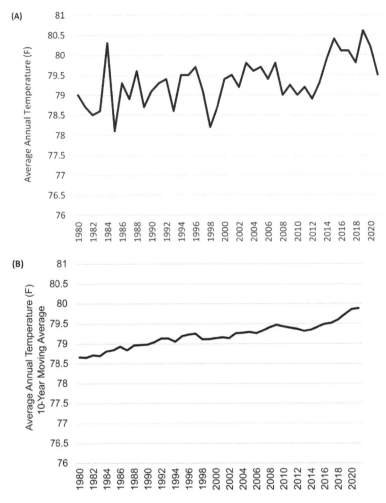

Figure 7.3 Make figures pleasing to the eye. (A) Lines like shark teeth are hard on the eye and obfuscate the takeaway. (B) A 10-year moving average grinds the shark's teeth and makes the takeaway clear.

risen over time. Replacing the annual temperature with the ten-year moving average grinds down the shark teeth and reveals the takeaway (fig. 7.3b).

Figures should also be simple. The best-laid-out figure will still be hard to grasp if it tries to convey too much. Simple figures have only as many elements as needed for one or two clear takeaways. "Housing bankruptcy rates are rising" has one takeaway; "household bankruptcy rates are rising, especially for young people" has two. "Household bankruptcy rates fell since the last recession but are now rising, especially for young people" has three takeaways and asks too much of the audience. Three-takeaway figures are for audiences of wonks where the purpose is to pore over data and for them to discover the takeaways for themselves.

Be proud of figures simple enough for a child to understand. Embrace a figure with only one line that clearly moves upward (as in figure 7.3b) or one with only two bars of very different heights. The overriding goal is to communicate the takeaway, not to make yourself appear sophisticated.

Last, the figure of the policy aide must be faithful, depicting a small difference as small and a large difference as large. The aide's purpose is not to create an emotional response, a certain vibe like that sought by a marketing graphic. This might be the goal of the lobbyist, the advocacy group, or the politician, but the aide's goal is to help the audience understand something worth understanding. You must therefore assess magnitude as described in chapter 5 and then build a figure faithful to the assessment's conclusion. The most pleasing and clear figure fails in its basic purpose if it leads the audience to think that something small is large. This is the essence of a misleading figure, and it can create as much misunderstanding as punching the wrong buttons on the calculator and reporting the wrong number!

One cause of an unfaithful figure is failing to think about magnitude at all. Only by chance would the resulting figure match what would have been your conclusion about magnitude. Another way to produce an unfaithful figure is to purposefully set the vertical axis so that a difference over time or across groups looks large (or small). If the rate of asthma in the population increases over time from 10.0 percent to 10.1 percent, the increase will look large if the vertical axis starts at 9.9 and ends at 10.2. A similarly misleading figure is one depicting percent changes for different groups with vastly different base values and therefore different absolute changes. Investment in producing hydrogen may have increased by 100 percent compared to a 5 percent growth in oil and gas sector investment, but in dollars the increase in oil and gas investment might be ten times the increase in hydrogen investment ($100 billion compared to $10 billion).

To further illustrate the need to assess magnitude when making figures, consider again the temperature data from the Honolulu air-

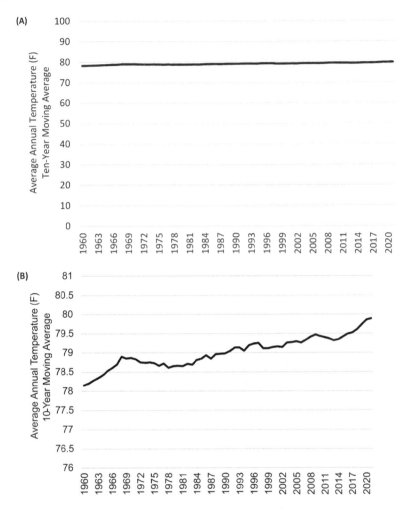

Figure 7.4 Two approaches to showing temperatures at the Honolulu airport, 1960–2021. (A) Starting the vertical axis at 0 degrees Fahrenheit. (B) Starting the vertical axis at 76 degrees Fahrenheit. Data are from NOAA National Centers for Environmental Information, Climate at a Glance: City Time Series.

port. You may have received the advice to always start the vertical axis at 0 and that doing anything else is playing games. But starting the axis at 0 degrees is no more objective than other possible starting points. For example, 0 degrees Fahrenheit is itself arbitrary and does not correspond to a lack of temperature. The axis range should of course cover the maximum and minimum temperature observed over the period, but how much space should be placed above and below them? More space makes any change over time look smaller; less space makes it look larger. The right amount of space depends on whether you assessed the magnitude of the change over time as large or small.

Figure 7.4 shows the ten-year moving average temperature, which has increased by 1.7 degrees since 1960. By itself, 1.7 degrees means little to most people. Setting the vertical axis to zero makes the increase nearly imperceptible (fig. 7.4a) and causes the audience to think that temperatures have been stable. But if your assessment of magnitude revealed that 1.7 degrees means that several species will go extinct, there will be a dramatic increase in the spread of invasive species, and crop yields will fall by 25 percent, you have reason to conclude that 1.7 degrees is large and create a figure that depicts it as such. Figure 7.4b, whose vertical axis starts at 76 degrees, is therefore the more faithful figure.

Some may object to creating figures to convey your conclusion about magnitude. The problem with the objection is that it presumes that some figures just show the data without conveying a particular sense of magnitude. But all figures convey a sense of magnitude. Figure 7.4a, whose vertical axis ranges from 0 to 100, conveys a sense of magnitude regardless of the intentions of the figure's author. The audience will have the impression that the temperature in Honolulu has been stable, when in fact it has increased enough to affect the island's ecology and economy. The aide's goal is to further understanding, but the supposedly just-show-the-data figure leads to misunderstanding rather than understanding.

Recall also from chapter 5 that any conclusion about magnitude should come with reasons for the conclusion. The point is not to manipulate the subconsciousness of the audience so they have a particular impression but lack understanding. When presenting figures orally or in writing, explain the reasons for labeling and presenting differences over time or across groups as large or small. Do not be surprised or discouraged if some people question your assessment of magnitude. If you did your assessment thoughtfully, you will have helped them think more carefully about magnitude even if their conclusion differs from yours.

8

Know How to Be Mostly Right (or at Least Respectably Wrong)

To be taken seriously by your audience and to further their understanding, work hard to be right. By working to be right, you will regularly be right, and if you happen to be wrong, you will be respectably wrong. As mentioned in chapter 1, by *right* I mean work that would not embarrass you if a statistically savvy journalist sifted every piece of it from start to finish.

This does not mean that your numbers are never wrong in hindsight. Discrepancies are respectable if you competently apply sound principles but reality turns out different from what your data and assumptions suggested. Discrepancies can occur when estimating the causal effect of what happened in the past, and they are the norm when projecting outcomes in the future like the number of COVID-19 cases over the next three months. Other discrepancies are less respectable and can arise from our own statistics or the interpretation and communication of the statistics of others. They include mistakes like using the wrong data, hitting the wrong key, misunderstanding units, and misinterpreting or misreading a statistic. The principles from the preceding chapters (e.g., know your data) will help you avoid some mistakes, and the principles from this chapter will help you avoid even more.

My appreciation for being right grew when I became responsible for generating official federal government statistics on farm household income shortly after finishing graduate school. The statistics of the US Department of Agriculture were official government statistics and had to be right. The secretary and numerous journalists might quote them within hours of being released, and once a wrong number starts circulating, it is hard to claw it back. Credibility suffers even if you do.

My experience at the Department of Agriculture was a helpful precursor to the Council of Economic Advisers. One of my first responsibilities upon arriving at the council was to learn its fact-checking process. Chairman Kevin Hassett expected a fact-check on every factual statement that reached him or that left the council. Every memo and slide started with a "Not Fact-Checked" heading that was removed only after fact-checking, which required that someone other than the author trace the path from the fact to its source.

Fact-checking covered facts from other sources, such as those drawn directly from an academic study, and facts established by the council, such as statistics calculated from original data. The path could be as short as following a link to an article to ensure that the fact described in the article matched the statement in the memo. Or it could be long and winding, involving many spreadsheets and lines of statistical software code. In confirming each factual statement, the fact-checker would insert queries where something was unclear or seemingly wrong. The statement's author then had to resolve all the queries to the satisfaction of the fact-checker before the memo could have the "Not Fact-Checked" heading removed and be sent to the chairman, the chief of staff, and whoever was the final audience.

Fact-checking absorbed many hours of council staff time, and not just that of interns and junior staff. Senior economists with PhDs had to do fact-checks when they required more technical expertise. And they did not just absorb the downtime between projects. The council often received an overwhelming number of requests for facts or analysis, some of which we could not fulfill because of limited staff availability.

So why did the chairman dedicate so much staff time to fact-checking? It made sense because increasing the council's output by 25 percent was pointless if it meant letting in error and losing credibility and effectiveness. The council would be more effective producing five requested memos a week than fifteen memos that no one requested or would read. The council's credibility, and therefore its ability to inform policy, rested on its work being right, or at least defensible, which is to say respectably wrong, if wrong at all. The chairman often had interviews with national media and needed to be confident that his statements would not return to haunt him. He also met frequently with the president, who might in a moment tweet a number to the world. Whatever bad numbers might come from the president or other administration officials, the council did not want to be their source. Its relevance depended upon its reputation.

Fact-checking is important even if you are not actually generating new facts. The closer a role is to a policy maker, the less it will involve generating new facts. The topics are too varied and too many; there just isn't the time. Instead, the policy aide very close to the policy maker focuses on filtering out unreliable and unhelpful numbers and interpreting good numbers for the context and audience at hand. Even here, fact-checking is important. Error will not stem from miscalculation—you are not calculating anything—but from misreading or misunderstanding a number. Remember, a number is much more than the digits that make it up. It is the story behind the number and its interpretation.

Enable Fact-Checking by Marking Your Trail

Only the simplest facts can be fact-checked without a clear trail provided by the author. The source for the statement "Median household income in 2020 was $67,521" is easy to find on the internet. But good luck checking something as simple as the result of multiplying two publicly available statistics such as household spending on gasoline (a Department of Labor statistic) and the projected change in gasoline prices (a Department of Energy statistic). Marking your trail can take many forms, and what makes the most sense depends on context, such as the software most used in your organization.

When drafting a memo or slide, insert comment bubbles with a link or file path to the source along with notes to guide the fact-checker to the appropriate table or paragraph where the fact is found. A statement about median household income in 2020 should have a comment with a link to the Census Bureau report and a note saying "See table 10." Even if no one else checks the facts, the comments will help your future self if you are pulled to other work for a week or more. The comment bubbles also allow you to spare busy people the clutter of footnotes, hyperlinks, and parenthetical citations. The clean final memo or slide deck should have a version with all the comment bubbles. Footnotes or endnotes can replace some bubble comments in longer pieces for broader external audiences, but even there it can be helpful to provide additional notes in comment bubbles in a separate version.

For facts that you or your colleagues generated, the comment bubbles should point to the document(s) with the calculations and the underlying sources. An Excel spreadsheet can be a useful place to document calculations and sources. Consider an example documenting several calculations made to answer a question about job creation associated with a policy recommendation.

As mentioned in chapter 2, in 2019 the US Department of Commerce shared a report with the White House recommending that the president require that US nuclear-power plants buy 25 percent of their uranium from US producers. Imagine that people involved in the White House policy process wanted to know how many uranium-mining production jobs the recommendation might create if adopted. One simple approach is to find the ratio of jobs to production during the last period when the US uranium-mining sector was expanding. Then multiply the ratio by the increase in production mandated by the policy recommendation. Figure 8.1 shows the Excel sheet documenting the calculations.

Fact-checking starts at a final number and follows the trail through any calculations and back to underlying external sources. Figure 8.1 therefore has only two types of numbers—those from external sources, such as government reports, and those calculated by me in Excel. Importantly, cells with calculations include only references to other cells. The calculation of the estimated increase in mining jobs

1	A	B	C	D	E	F
2	Item	Value	Units	Source	Notes	Link
3	2018 uranium production	0.72	million pounds	2021 Domestic Uranium Report	Table 2	www.eia...
4	2012 uranium production	4.34	million pounds	2021 Domestic Uranium Report	Table 2	www.eia...
5	2012 uranium production jobs	1196	jobs	2021 Domestic Uranium Report	Table 6	www.eia...
6	jobs-production ratio	276	jobs / million pounds	calculation	jobs / production	
7	Commerce Dept. production target	6	million pounds	Commerce Dept 232 report	Page 41544	www.govinfo...
8	estimated increase in mining jobs	1456	jobs	calculation	(prod. growth) x (job/prod ratio)	

"=B6*(B7-B3)"   "=B5/B4"

Figure 8.1 Example documentation for estimating job creation from uranium production

appears as "B6*(B7-B3)" instead of "276*(6-0.72)". The expression "276*(6-0.72)" is a dead end, whereas "B6*(B7-B3)" points to places with a description of the numbers, their units, and their sources.

Imagine if one of the numbers lacked documentation, having been initially produced by an intern who is back in college and is not responding to email. Depending on what other information the sheet has and the obscurity of the number, it might be impossible to confirm that the number is appropriate. This means no source, no description, and no units for the number. Good luck proving to a critic that the number and any that depended on it are correct.

Not every calculation must occur in the same Excel sheet. The key is having a well-marked trail from the final number through any calculations and ultimately arriving back to numbers from external sources such as a report or data set. The sheet with the final number, the parent sheet, may reference a cell in a child sheet that has a statistic calculated from data in the child sheet. The child sheet, in turn, should have information on the source of its data. Likewise, more substantial data manipulations and calculations may occur outside Excel entirely, but where the resulting numbers appear in the parent sheet, there should be pointers to the file that the statistical software ran to produce them, such as a script file for R or a do-file for Stata. The script file should read in the original data and have notes about the source and the calculations and manipulations that occur in it. This ensures that the trail goes all the way back to the original sources.

For more complicated projects that require combining the output from numerous data sets, documenting the big picture helps you and the fact-checker. The documentation could take the form of a diagram in a separate Excel sheet where you use boxes and arrows to describe the data sets, their output, and the relationships among them. The fact-checker or your future self will thank you for the documentation.

Have Your Numbers Fact-Checked, Preferably Not by You

It is easy to make mistakes, especially ones that make you embarrassingly wrong and not respectably wrong. We can read 6,325 instead of

6,235; we can assume that "tons" meant short tons instead of metric tons (one is 10 percent heavier than the other); and we can mistakenly calculate $(25 + 3) \cdot 2$ instead of $25 + (3 \cdot 2)$. Even if no one is available to fact-check the work at the moment, assume that someone will do it eventually, and that person might not be friendly to the claims supported by your number. So when you are rushing to arrive at a final number, document it as if someone else who has little familiarity with the topic, the variables, or the units will check it. This may be your future self when an expert on the topic or a diligent reporter calls you and wants to go through your calculations.

Fact-checking can occur at a shallow or deep level. The calculations in figure 8.1 suggest that adopting the Commerce Department's recommendation might create about 1,500 jobs in uranium mining. A shallow fact-check ensures that another person can replicate the work, arrive at the same final number, and agree with its interpretation. It involves the value of numbers and their meaning. Consider that US uranium production in 2018 was 0.72 million pounds. Is the number really 0.72? Is this really US uranium production? For 2018? In millions of pounds? The description of the work also says that the jobs-to-production ratio was multiplied by the increase in production. Was the production-to-jobs ratio mistakenly used?

A deep fact-check considers the conceptual soundness behind the numbers and calculations. To pass a deep check, the methods and assumptions do not have to be the most sophisticated or precise. Rather, they must be reasonable and give a sufficient foundation for the precision needed for the purpose and audience at hand. Is it reasonable to multiply the jobs-to-production ratio by the increase in production? This assumes that each unit of production requires a certain amount of labor and that the amount is similar at different scales of production. And how reasonable is reasonable? If the goal is to know if the growth in the uranium-mining jobs is closer to 1,000 or to 20,000, the ratio approach is reasonable. It should not be off by a factor of 10. But if it matters that it is 1,500 instead of 1,000, a more sophisticated approach might be warranted.

In addition to occurring at difference levels (shallow versus deep), fact-checking can occur to different degrees. A complete shallow fact-check means replicating all the data and statistical work needed to arrive at the final number—a literal redownloading of the data and replicating of all the steps to ensure that the same numbers emerge. Likewise, a complete deep fact-check means putting under a magnifying glass many statistical or mathematical modeling decisions and assumptions. Alternatively, only the most essential calculations or assumptions could be scrutinized.

The different levels and degrees of fact-checking raise a practical question—how much fact-checking effort makes sense? In their book *A Practical Guide for Policy Analysis*, Eugene Bardach and Eric Patashnik remind the policy analyst that it is costly to produce evidence. When considering which evidence to gather, the analyst should consider its value, such as whether it could lead to a different decision.[1] A similar principle applies to fact-checking. Is a more thorough check worth the reduced risk of being wrong? The cost of being wrong can vary, being low in situations where the statistics are just for context and high in situations where the statistics are the hinge that swings the policy one way or the other.

A related question is how much uncertainty can surround a number before it becomes worse than no number. One measure of a good or a bad number is simply whether it helps a desired end such as garnering media attention or securing votes. A better measure that the policy aide should use is whether the number fosters an understanding that matches reality. By this measure, some numbers can be too uncertain to use because they will likely shift understanding further from reality, not closer to it. This could be in cases where the standard error is very large, indicating that a different sample could produce a wildly different number and an entirely different understanding of the issue. Another case is where a number arises out of layers and layers of poorly vetted assumptions, and the violation of any one of them could yield a very different number.

Navigate the Gray of Fact-Checking with Care

Fact-checking can be straightforward. Is this data set described correctly, and is 6.2 percent the sample average value? But a fact to one person may be disinformation to another. Disagreements about facts are not limited to distrusting partisan foes. Some lively discussions within the Council of Economic Advisers centered on whether a statement passed the fact-check bar; this was among people who respected one another and worked well together. Four major gray areas in fact-checking involve model projections, causal effects, dated facts, and assessments of magnitude.

Model Projections

Model projections of what might happen in a future scenario are common in policy discussions. The models draw from theory and statistics to simulate a part of the world, such as its climate, its economy, or the health of its people. For example, complex climate models combine historical data with knowledge of earth systems (e.g., cloud formation) to project temperature trends under different scenarios of greenhouse-gas concentrations. Likewise, models of the economy project the employment rate under different interest rate policies, and epidemiological models project infections and deaths under different scenarios of government policy and human behavior.

Much gray in reporting model projections is in how much confidence to place in them. Are these things that will happen, are likely to happen, or that might happen? Quantifying uncertainty is hard even for people who live and breathe the details of the models. Two communication tips, however, can avoid much misunderstanding.

- *Projections are projections.* Make sure that the audience understands that a projection is just that, a projection, or what might also be labeled a forecast, a prediction, or an estimate. This reminds people that the numbers are about a future that no one foresees perfectly.
- *Projections correspond to scenarios.* All projections are associated with a particular scenario, which is a package of assumptions about how the world does or will work. Always keep the projection wedded to its scenario as described by the original authors. This helps the audience consider the plausibility of the projections for themselves.

If practiced broadly, the tips could add much clarity to public understanding and policy discussions. Consider the report from March 16, 2020, in which researchers at the Imperial College of London projected COVID-19 deaths in different scenarios.[2] The death projections for the summer of 2020 quickly entered the public conversation and internal policy debates, including the White House, and were sometimes miscommunicated and misunderstood.

Media reporting about the projection generally used the word *projection* or *prediction* or inserted the word *could*. This is important because there is a vast difference between "We predict that 2.2 million people in the US would die of COVID-19 absent any control measures by governments or behavioral changes by individuals" and its shortened version "2.2 million people in the US would die absent any control measures by governments or behavioral changes by individuals." Despite its vastness, the difference can easily be lost in discussions and presentations as it was in a March 31 White House briefing in which the Imperial College report projections were depicted in a figure and labeled as "Pandemic Outbreak: No Interventions." There is not a *projected*, *predicted*, or *forecast* in sight.[3] Too much was sacrificed for conciseness.

Moreover, informal and formal reporting on the projections often did a poor job of wedding them with their assumed scenario. The actual wording of the Imperial College report was "In the (unlikely) absence of any control measures or spontaneous changes in individual behaviour . . . we would predict approximately 510,000 deaths in [Great Britain] and 2.2 million in the US." The following day the *New York Times* described the projection as reflecting what could happen from an "uncontrolled spread" of the disease.[4] In the White House briefing, the 2.2 million deaths were labeled the "No Intervention" case. In both cases, a *Times* reader or an attendee at the briefing could naturally understand the projection to reflect a scenario of no control measures by government (the *Times* wording) or no intervention by government (the White House wording).

Thus, people began thinking that absent government lockdowns or mandates, 2.2 million people are likely to die soon. What is missing is that the scenario assumed more than no government intervention to control the virus. It also assumed no "changes in individual behavior." The projection is for a scenario in which people are surrounded by news of infections and footage of morgues filling with bodies and yet never put on a mask or avoid a crowded room. The authors of the projections explicitly stated that this was unlikely, and data available at the time also showed large changes in people's behavior apart from government mandates. A 2021 study of the early stage of the pandemic found that "legal shutdown orders account for only a modest share of the massive changes to consumer behavior."[5]

Keeping the projection wedded to its scenario would have prevented the misunderstanding that the projections were the best guess of what would happen absent government intervention. It would have allowed people to better gauge the plausibility of the projection. Although it is hard or impossible for most of us to step inside the details of a complex epidemiological model, anyone can ask, "If I hear of hospitals filling with people dying of an infectious disease, would I do anything to reduce the risk of infection?"

Causal Effects

Another gray area for checking facts regards statistics alluding to or indicating causal effects. "Correlation does not necessarily imply causation," says your statistics text and professor as they mention the example ice cream sales and murder rates. Clearly it is a wrong to speak of ice cream vendors as having blood on their hands. Most cases, however, are less clear, and the mantra about correlation not implying causation says only that we should not blindly interpret correlations as causations. It gives no guidance to navigate cases where there is a correlation and a plausible channel between the cause and its supposed effect (something missing from the case of ice cream sales and murder rates).

Two tips aid in handling estimates that might be interpreted as causal effects.

- *Defer to the authors in cases of association.* If the authors speak of their statistic as an association or a correlation, do the same unless you are knowledgeable of the topic and confident enough to absorb the details of the research design and find good reason to make a stronger claim. Some disciplines use the language of association or correlation for almost any observational study despite the rigor of its research design.[6]
- *Apply basic tests in cases of causal claims.* If the authors speak of their statistic as reflecting a causal effect, subject the claim to a basic plausibility test and a basic research design test. For the plausibility test,

are the causal channels and the magnitude of the causal effect plausible? This test catches the ice cream–murder case. It also catches cases where a $10-per-student educational grant improves standardized test scores by 50 percent. The basic research design test looks for the four main parts of a credible research design described in chapter 6: a change part, a cross-sectional part, a temporal part, and a counterfactual part. A lack of one of these parts generally weakens the claim for causality.

To apply the tips, consider a study by researchers at the Johns Hopkins Bloomberg School of Public Health. The researchers studied the predictors of indoor concentrations of radon, a carcinogenic gas, in Pennsylvania homes. They found a "statistically significant association between proximity to unconventional natural-gas wells drilled in the Marcellus shale and first-floor radon concentration." The authors are consistent in speaking of their result as an association. In covering the study, the NBC News headline was "Rising Levels of Toxic Gas Found in Homes Near Fracking Sites" and reported that the trend in rising values over time "was linked with just how much unconventional drilling was going on."[7]

Would the NBC statement about the study's findings pass your fact-check? Applying my tips, the study authors likely chose the term *association* to convey their lack of confidence in a causal interpretation of their statistic. The NBC article uses language that conveys clearer claims of causality than the authors make. Reporting that rising radon values were "linked" to drilling suggests a causal effect—it evokes the imagery of a chain or rope linking two different things. Links are not mirages of connectivity but actual connections.

But perhaps there is a basis for a causal claim, so let's apply plausibility and research design tests. It is plausible that unconventional natural-gas drilling (fracking) occasionally opens pathways for radon in the subsurface to reach the surface and perhaps into homes. The probabilities might be low, but it's not an absurd scenario. The study, however, lacks the temporal component of a credible research design, which involves having data from before the change in the thing whose effect is under study. The study links the number of wells already drilled around a home with the home's current radon level, not the change from its pre-drilling radon level. This lack of a pre-drilling measure of radon weakens the claim for a causal effect. In fact, my colleagues and I accessed the same data as the study and found that the statistical association between wells drilled and indoor radon disappears when using a more rigorous approach that allows wells to be located in areas that already had higher radon levels.[8]

Dated Facts

A third area of grayness is in communicating dated facts that may not have aged well. They require additional detective work that searches for more recent statistics or articles that may cast a new light on the facts. Otherwise, you risk giving the audience a cutting-edge understanding from a decade ago. This could bite you or them when engaging with someone with a more current understanding.

New information can substantially revise the understanding of an issue, such as the evidence for a causal relationship. The radon well example illustrates the point. If someone in 2022 adopts the NBC article language that indoor radon concentrations are linked to drilling gas wells, they are communicating an outdated understanding. The more recent study by myself and my coauthors used the same data but more rigorous methods and found that the statistical association was spurious. Of course, a new study in a different context might discover other issues surrounding gas drilling and radon, perhaps requiring a qualification of our study.

Recall also the *Diamond Princess* example from chapter 2. Statistics on COVID-19 were so limited at the time that those from the *Diamond Princess* improved the understanding of presidents, prime ministers, and health officials around the world. In early March 2020, the statement "the best evidence suggests that COVID-19 has an infection fatality rate of roughly 0.15 percent" should pass a shallow and deep fact-check.[9] But after several billion infections, we should no longer rely on the *Diamond Princess* sample of seven hundred infected people. It turns out that this infection fatality rate aged well. A year later, a survey of many studies indicated an average global infection fatality rate of around 0.15 percent.[10] But the early and late estimates of the fatality rate could have been quite different.

Assessments of Magnitude

Last, assessments of magnitude also lie in the gray of fact-checking. A report or article may speak of a statistic as ignorable or disconcerting without clear reason. You may spread misunderstanding if you blindly adopt the assessment of others. Instead, follow the guidance from chapter 5 for discerning large and small problems and effects. Convince yourself and a friend that your adjectives are defensible. Suppose, for example, that the authors of the drilling and radon study had described their estimated association as large. The actual estimate was that one well drilled per square kilometer was associated with a 2.8 percent increase in indoor radon concentrations (or a 1.7 percent increase for the median home in areas with drilling, which had 0.6 wells per square kilometer). If this association is large, then other associations presented in the study are gigantic: radon concentrations were 21 percent higher in buildings dependent on well water and more than 200 percent higher in homes in the Nittany formation than in the Stockton formation.

Be Aware of the Pressures upon You

Our leanings or those of our audience surface in gray areas and can distort our communication of statistical information. Would the NBC statement about rising radon being "linked with" drilling pass your fact-check bar? If we place a high value on reducing fossil-fuel use, we will be more inclined to interpret the association as causal even though our preference against fossil fuel is unrelated to the methodological considerations that would support (or not) a causal interpretation. Even if we are agnostic about natural-gas drilling, our audience, perhaps an environmental nonprofit, might seek to reduce natural-gas drilling. The pressure will be toward an interpretation that pleases the audience, especially if it includes our direct boss or client.

Fight to be aware of the pressures upon you—whether from personal leanings or those of important people around us—and seek to provide interpretations that can withstand fact-checking by disinterested parties. It might not excite your audience as much, but being careful with interpretation and quality of evidence will build your credibility and guard against misunderstanding in your audience.

The Fruit of Fact-Checking (and This Book)

A goal of this book is to build knowledge held confidently enough to be wielded in situations where health, money, or reputation are on the line. Fact-checking and the documentation that it requires contributes to this end by cultivating a habit of carefulness and thoroughness. The habit has its fruits. It will grow your confidence that your numbers are right, or if wrong, respectably wrong. In time, it will also show yourself and your statistics to be credible and therefore to be taken seriously. And because your numbers are often right, you will be giving the people you serve a better map as they navigate the challenges of policy making.

Indeed, every part of this book has been designed to help you grasp essential principles that if you get wrong, none of the details will matter. The principles help you make statistical maps that are properly scaled and that put major highways and towns in proper relation to one another. I hope that it has succeeded.

9

Fail the Test? A Case Study in Using Statistics for Policy

SHOULD THE UNIVERSITY OF CALIFORNIA
USE STANDARDIZED TEST SCORES IN ADMISSIONS?

Obtaining a driver's license often requires passing a knowledge exam and a road exam. The knowledge exam asks about traffic laws and driving practices; the road exam has the student drive a real car on a real road with real traffic. One can score well in knowledge and yet fail the road exam, as I did when I was sixteen. The discrepancy in performance arises because reading books about driving is not driving. A driver education class cannot prepare you to gauge the distance between you and the stoplight and apply the brakes gently enough to avoid a crash from behind and hard enough to stop before entering the intersection. This can be learned only through practice—on flat roads, downward-sloping roads, upward-sloping roads, in the day, at night, in sun, in rain, and in snow. To use terms from the introduction: not all knowledge is wielding knowledge, which is knowledge held firmly enough to be used confidently in decisions with real-world consequences.

Wielding knowledge of statistics comes through selecting and interpreting statistics in diverse real-world situations, which is why prior chapters reference many examples. But the examples thus far have been brief and used only to illustrate a point or two. None follow an analysis from beginning to end to unpack the statistical concepts and judgments employed along the way. This is the thrust of this last chapter. It is a case study on using statistics to inform a policy decision with immediate and lasting consequences for many. It is not actual driving, but it is the next best thing. Think of it as entering a simulator with pedals, a steering wheel, and a screen showing a road with cars passing by. The case study illustrates various points of the book and will be helpful even if you are not interested in education policy, just as practicing driving is helpful even if it is on a road you will never again travel.

The Policy Maker's Request

In 2018 the president of the University of California, Janet Napolitano, asked the university's Academic Senate "to examine the current use of standardized testing for admission to the University of California . . . and determine whether any changes in policies on the use of test scores are needed."[1]

The senate then created the Standardized Testing Task Force to fulfill the charge. It had the role of the policy aide described in chapter 1. As with the aide, the task force did not have the authority to make a decision, nor was it meant to approach the charge with a prior commitment to a particular outcome. It was created to further the understanding of the policy maker—in this case, university leadership—and provide it with a recommendation that it would be free to adopt, modify, or reject.

President Napolitano did not request a mere statistical report. To provide one would be to follow the example of the auto mechanic in chapter 5 who just wanted to measure the width of the tread on your tires instead of helping you decide whether new tires were needed. That would not do. The president asked the task force to assess how current practice serves the university and its students, and whether an alternative arrangement might serve it better. This is more than crunching every statistic imaginable. It is carefully selecting, calculating, and interpreting statistics so as to illuminate a problem and a path forward.

The University of California has roughly half a million students, faculty, and staff, so its president is a busy person. Why did she divert her attention from ten other matters and direct it to the use of standardized tests in admission decisions? President Napolitano's request is not publicly available, but the task force report lays out the high-level problem and a possible cause: in 2019, underrepresented minority students accounted for 59 percent of California's high school graduates, but only 37 percent of resident students admitted to the freshman class. At the same time, underrepresented high school students in the state had considerably lower average standardized test scores than White or Asian students in the same year, scores that the university incorporates into admission decisions.[2] In the language of chapter 5, the statistics indicated a large problem.

Do you see the role that well-selected simple statistics play? The comparisons in enrollment rates and in test scores were likely influential, if not decisive, in capturing President Napolitano's attention. This fits the *Should I Care?* situation in which statistics can be decisive for capturing the scarce attention of policy makers (chapter 1). Of course, someone had to package them in an argument and force the issue; statistics do not by themselves come to the ears or eyes of busy people. But the ability of a group to spur action is less if all it can do is tell anecdotes about how this person's neighbor or that person's nephew was not admitted despite good grades in high school. The statistics, in contrast, highlight a systemwide issue and rebut the argument that the anecdotes are rare cases.

Having Multiple Policy Goals Calls for More Work

But why did the disparities in enrollment and test scores cause President Napolitano to ask for a study instead of making an immediate policy change ending the use of test scores? If the university cared only about having a freshman class whose racial composition reflected that of California's high school seniors, the simple statistics and the disparities they show may have been sufficient for action. But if decreasing disparities were the only goal in play, the university could replace its admission staff with a computer that randomly selects seniors for admission without regard to any of their characteristics (and then provide aid based on need).

There is a good reason why the university has not replaced its admission staff with a computer. The university's admission policy pursues multiple goals: "The University seeks to enroll, on each of its campuses, a student body that, beyond meeting the University's eligibility requirements, demonstrates high academic achievement or exceptional personal talent, and that encompasses the broad diversity of cultural, racial, geographic, and socioeconomic backgrounds characteristic of California."[3]

In other words, the university values excellence ("high academic achievement or exceptional personal talent") and having a student body representative of the state's residents. With the additional goal of excellence in play, the best path forward is less clear. A few simple statistics were influential and perhaps decisive for capturing the president's attention but not decisive for her to change policy. The decision involves understanding how standardized tests affect several goals that may compete with each other. Indeed, the task force stated that "it assessed how to weigh the goal of accurate prediction of UC performance against other, potentially competing goals, such as educating future leaders, representing the diversity of California, promoting socioeconomic mobility, and countering inequality."[4]

The competition among goals complicates the policy decision, and it is why a few simple statistics do not resolve the issue. More work is needed that requires the detective and statistical skills mentioned in chapter 1. The detective absorbs the particulars of a situation and pieces together an account of what happened or is happening. In the case at hand, fulfilling the request requires an understanding of the university's values and goals. How can you assess "whether the university and its students are best served by UC's current testing practices" if you have no sense of what it means to best serve the university and its students?

The statistician is of course needed too. How do you measure the predictive power of test scores after accounting for high school grade-point average (GPA)? On what sample of students and over what years should you estimate predictive power? How might the finding apply to students in coming years? It turns out that the university task force did much statistical and detective work, resulting in 228 pages "grounded in evidence-based research and UC values."[5]

Understanding and Answering Key Empirical Questions

The task force's report addressed many dimensions of a complex issue, and this case study does not convey its breadth nor reassess its conclusions. Instead, it highlights a few parts of the report that illustrate important judgments in selecting and interpreting statistics to inform a policy decision.

Before diving in, recall from chapter 2 that learning from data is harder than it sounds. It requires going in with a commitment to listening to the data on the basis of its relevance and quality, not on whether it says what we expected or like. It appears that the task force had this posture. It openly noted being surprised by the data, that some members initially thought that lower average scores for underrepresented groups played a large role in admission disparities but through careful study learned otherwise.[6] The statistics "showed, unexpectedly, that test scores likely make a surprisingly modest contribution to [disparities in admissions]."[7] Being surprised and admitting it is a sign of a healthy ability to learn from data.

Do Standardized Test Scores Predict College Success?

A key empirical question addressed by the task force is whether standardized test scores help admission staff identify students likely to succeed at the University of California. If we compare two applicants with the same high school GPA, does knowing their test scores help us select the one more likely to succeed in class and graduate? The answer has much bearing on the policy question at hand. If test scores do not aid in selection, no goal suffers from abandoning them. The task force can write a short report with a simple recommendation: drop the tests now.

But if they do aid in selection, dropping them would reduce the university's ability to identify the most promising students. By extension it would select more students less likely to succeed (assuming that the university does not shift attention to other information uncorrelated with high school GPA but as correlated with success). It would not serve the university or its students to change policy in a way that increases the number of students spending time and money at the university but who leave without a degree.

The empirical question is not about the causal effect of test scores but about their predictive power. We are not interested in an experiment in which scores are randomly assigned to students and their causal effect (perhaps through some psychological boost) on college success estimated. We want to know if scores help admission staff select students more likely to succeed such that the mix of students admitted using scores (and everything else) graduate at a higher rate than those admitted without using scores (but using everything else). This is an issue of predictive power. In the language of scatterplots, are the data wildly scattered around the score-based best-fit line, or do they hug it closely?

MEASURING PREDICTIVE POWER AND DISTINGUISHING LARGE FROM SMALL

To assess predictive power, one needs something to predict. Since its client was the university, the task force used four outcomes that the university typically uses to track academic success: retention to second year, first-year GPA, graduation (on time or not), and GPA at graduation. Next, it needed to measure the predictive power of test scores after accounting for high school GPA. This is the "accounting for X" use of regression described in chapter 4.

The authors compared the R-squared of two regressions, one in which high school GPA is the only independent variable and one in which both GPA and test scores are included. Note that a regression with two independent variables is like a regression of the outcome (e.g., first-year college GPA) on the test score variable after it has been purged of the part of its variation predicted by high school GPA. This is what was done to isolate the relationship between road spending and road quality apart from neighborhood income in chapter 4, except that the three variables have been replaced, respectively, by first-year college GPA (the outcome), test scores (the variable of interest), and high school GPA (the variable to be accounted for).

Though not discussed in the report, the task force estimated predictive power without adding other applicant characteristics such as race or household income. They could have included them, which would change the interpretation of the estimates of predictive power. The regressions would give the predictive power of the variation in test scores uncorrelated with high school GPA and everything else included in the regression. Practically, this raises the question of how the additional factors should be included (e.g., family income as a quadratic function or as a series of dummy variables). Conceptually, should we think of admission staff as perfectly holding in their mind all the various characteristics of an individual and then comparing students based on test scores? Or should we think of them as holding high school GPA in mind and perhaps one other characteristic, such as a student's family income, and then comparing on the basis of test scores? The task force implicitly took the second view, controlling for only high school GPA but estimating predictive power within each group of students (e.g., income groups, race/ethnicity groups). One strength of the approach is that it shows if the predictive power of test scores is greater for some groups than for others.

Before looking at the results, what would it look like to have little predictive power or a lot of predictive power? This is the difficult problem of knowing large from small (chapter 5). The task force addressed the problem with seriousness because its members knew that they were writing not to an amorphous academic community but to university leadership that needed help in making a decision that would affect hundreds of thousands of people.

On the way to an answer, the task force presented a common rule for classifying predictive power, where low predictive power is an R-squared less than 0.1, moderate power is between 0.1 and 0.25, and high power is greater than 0.25. It then proved a point from chapter 5—that contextless rules to distinguish large from small have limited value in practical settings. The task force noted the rule and then quickly abandoned it, stating, "Even a low correlation coefficient can reveal large differences between the expected success levels for students in different grade or score bands. . . . The practical usefulness of a given strength of correlation depends on the fraction of applicants who can be selected."[8] If few applicants are admitted, even weak correlations can yield a large increase in the proportion of students who will do well.[9] In short, the predictive power of a variable should not be called low simply because it has an R-squared of less than 0.1. The variable might increase the predictive power of a model by 50 percent and the number of students graduating by thousands.

DATA AND SAMPLE

The task force assessed how well test scores predict college success using data from the freshman cohorts of 2010 to 2012, which was the most recent cohort with data on all outcomes of interest (including seven-year graduation rates). So, is this a random sample? Clearly not. It includes only students who applied to the University of California and were selected for admission and actually enrolled. But remember from chapter 2 that "Is this a random sample?" is the wrong question. Instead, ask the three questions from chapter 2 about origins, purpose, and generalizability.

First, how does someone enter the sample, and who are they as a group? They enter by applying, being admitted, and then enrolling. As such, that person is a higher academic performer on the whole than the general population of high school students and, given the selectivity of the University of California, perhaps a higher performer on average than the average student applying to other California four-year institutions.

Second, what do we want to learn from the sample? We want to learn how well standardized test scores predict college success for the purpose of potentially changing admission policy for future applicant pools. As such, we are interested only in the 2010–2012 cohorts to the extent that they tell us the predictive power of test scores in the coming years. This is probably why the task force did not include earlier cohorts (e.g., what about the class of 1980?). Yet even the 2010–2012 cohorts have long left the university, which raises the next question.

Third, do sample statistics apply to other groups? This relates to the prior question of purpose and the suitability of the cohorts. Statistics from the 2010–2012 cohorts probably apply well enough in the short term, absent major changes in the profile of applicants or admits. That said, the task force noted a trend in increasing high school GPAs, which is known as grade inflation.[10] We might therefore expect the predictive power of test scores to increase with future cohorts, particularly in comparison to the predictive power of the baseline model with only high school GPA.

THE FINDINGS

The first column in table 9.1 shows the R-squared from a regression using high school GPA to predict each student's first-year college GPA; the second column shows the R-squared when the SAT variable is added to the regression. Recall that the R-squared ranges

Table 9.1 SAT Scores Help Predict Freshman GPA apart from High School GPA

| Student group | Regression R-squared | | Change | Percentage change |
	High school GPA only	Adding SAT scores		
All	0.16	0.26	0.10	63
<$30K	0.12	0.22	0.10	83
$30K–$60K	0.13	0.23	0.10	77
$60K–$120K	0.13	0.20	0.07	54
$120K+	0.17	0.20	0.03	18
Asian	0.15	0.24	0.09	60
Black	0.09	0.17	0.08	89
Hispanic	0.11	0.20	0.09	82
White	0.14	0.18	0.04	29
Not first generation	0.15	0.21	0.06	40
First generation	0.12	0.21	0.09	75

Note: The statistics reflect the 2010–2012 cohorts of students admitted and enrolled in the University of California and are drawn from the STTF Report's table 3A-1 (third column) and table 3A-2 (fourth column). They are the R-squared values for a linear regression of freshman GPA on high school GPA only and a regression of freshman GPA on high school GPA and SAT scores.

from 0 to 1, where a value of 1 indicates that all the variation in one variable is predicted by variation in the other variable. In scatterplot terms, the data points would fall perfectly on the best-fit line with an R-squared of 1.

For the full sample, adding SAT scores to the regression increases the R-squared by 0.10, which is considered low to moderate predictive power according to the standard contextless classification rule. But in proportional terms, adding SAT scores increases the R-squared by 63 percent.

Breaking out the analysis by various groups reveals that SAT scores add more predictive power for students in underrepresented groups. For example, it increases power by 89 percent for Black students compared to 29 percent for White students. For students from households making less than $30,000, the increase is 83 percent compared to 18 percent for those from households making over $120,000. It is not hard to think of reasons for the difference. The high scores of high-income students are more likely to reflect access to tutors or special SAT classes, which may correlate poorly with freshman GPA. By comparison, a high score for a low-income student is more likely to reflect exceptional ability and strongly correlate with freshman GPA. The task force does not leave us to speculate on the practical consequence of the predictive power of SAT scores. To complement the R-squared comparisons, it broke students into groups based on their high school GPA and their SAT score (for GPA, 3.00 to 3.25, 3.25 to 3.5, and so on; for SATs, less than 850, 851–1000, and so on). Within each GPA group, they compared measures of college success for students with low SAT scores to those with high scores. They found this: "For any given high school GPA, a student admitted with a low SAT score is between two and five times more likely to drop out after one year, and up to three times less likely to complete their degree compared to a student with a high score."[11]

To appreciate the finding, consider two students with a similar GPA and otherwise similar records. The finding means that knowing the students' SAT scores aids in identifying the one who is more likely to graduate. And if the difference in SAT scores is large, the difference in the likelihood of graduating will also be large. As with the R-squared analysis, breakouts reveal larger differences in outcomes across high- and low-score students in underrepresented groups (again, comparing within the same GPA group). So, among the two underrepresented students with similar GPAs, test scores are even more helpful in selecting the one most likely to graduate.[12]

So How Are Test Scores Used in Admissions?

The finding that test scores predict college success after accounting for high school GPA primarily concerns the university's goal of academic excellence. What about the goal of representing the diversity of California residents? After all, the task force was formed in part out of concern that using test scores in admission thwarts progress on the representation goal. Given the lower average scores of underrepresented groups, this would occur if the university used hard and universal admission rules based on standardized test scores. This is not the stated practice, which is to comprehensively review applications on the basis of fourteen factors.[13]

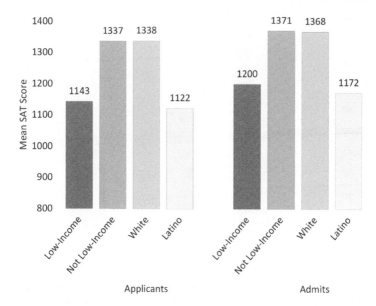

Figure 9.1 Differences in group SAT scores appear among applicants and admits. The statistics reflect 2019 applicants and admits to the University of California and are drawn from the STTF Report's table 3C-4.

How the comprehensive review incorporates test scores is unclear. Simple mean comparisons, however, shed light on how it plays out in practice. If test scores receive a dominant weight in admission, differences in scores across groups of applicants should largely disappear when looking only at admitted students. Low-scoring students would generally not receive admission, so most admitted students would have high scores regardless of their racial or income group. For this exploration the task force used the most recent year of admission data (2019) and not the earlier data used to assess the power of test scores to predict later performance in college. This reflects its preference for the most recent data as likely most informative for a forward-looking policy decision.

The 2019 admission data show large differences in mean test scores among groups of applicants (the left group of bars in figure 9.1) and also large differences across groups of admits (the right group of bars). For example, low-income applicants have much lower scores than other students, and this is also true among admits. Comparing White and Latino students shows a similar pattern.[14] Admits have

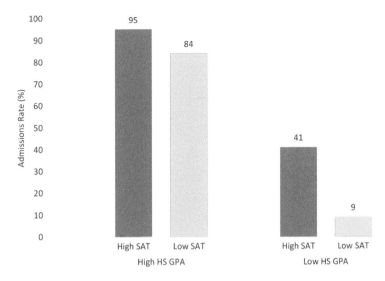

Figure 9.2 SAT scores matter most for selecting among low-GPA students. The data come from the STTF Report's table 3B-1 and are based on 2018 admission data from the University of California. A low GPA or SAT is defined as being one or more standard deviations below the mean for California resident applicants; a high GPA or SAT is one that is one or more standard deviations above the mean.

higher mean scores than applicants, so scores clearly receive some weight by the university, but it is used to select among students within the same group. Among two low-income students, for example, the one with a higher score is more likely to be selected, but low-income students do not need scores similar to other students to gain admission. If that were true, low-income admits and other admits would have similar mean scores. As it is, low-income admits have a mean score 171 points less than other admits, only slightly smaller than the 194-point difference observed among applicants.

Related admission statistics reveal that low SAT scores are unlikely to keep strong high school students from being admitted to a University of California campus. As shown in figure 9.2, going from a high SAT score to a low score reduces a student's chance of admission from 95 percent to 84 percent if she has a high GPA ("low" is defined as being one or more standard deviations below the mean for California resident applicants, and "high" is being one or more standard deviations above the mean). In contrast, a low high school GPA greatly reduces the chance of admission. Even with a high SAT, an applicant with a low GPA has only a 41 percent chance at admission. If a student has a low GPA and SAT score, the admission rate falls to 9 percent, indicating considerable weight being given to SAT scores when selecting among low-GPA students.[15]

What Is Happening to Minority High School Seniors?

Like a good policy aide, the task force was not satisfied with gaining a clear insight into only one corner of the problem. It sought answers to all the major questions. If the comprehensive review is primarily using test scores to select from within applicant groups, as the data suggest, what explains the statistics cited in the beginning showing that underrepresented minorities make up a much larger share of the high school senior population than they do of admitted students to the university? To answer, the task force looked at the population of California high school seniors who were underrepresented minorities and not admitted to the university in 2017 or 2018. They traced the senior's next steps to see what kept them from admission.

First, 10 percent of high school seniors did not graduate, and a further 40 percent did not complete the high school courses required by the university. The remaining 50 percent of students are split evenly between those who did not apply and those who applied but were not admitted.[16] Standardized test scores play a role only in the last half because the tests should not affect which courses students complete, nor should their likelihood of graduating. It probably affects whether some high school students apply because standardized tests cost money, and for those who do take them, a low score could discourage them from applying. And of course test scores account for some portion of the remaining quarter of students who apply and are not admitted.

On the whole, the breakout indicates that the greatest barrier to admission over which the university has control are high school course requirements. Assuming that test requirements do not have very large effects on the propensity for a student to apply, the next greatest barrier over which the university has control involves high school GPA. The statistics from the prior section show that it likely accounts for a greater portion of the last 25 percent (applied but not accepted) than do standardized test scores: a high-GPA student has an 84 percent chance of admission despite a low SAT score, but a low-GPA student has only a 41 percent chance of admission despite a high SAT score.

The Recommendation and the Decision

On the basis of its careful study, only some of which is reported here, the task force made numerous recommendations. In particular, it recommended that the university not make standardized tests optional for applicants as it "might make it *harder* . . . rather than easier for UC to make up disparities due to the other factors."[17]

At the same time, the task force acknowledged that its statistical work was more of a helpful map that laid out various paths and destinations rather than an irrefutable destination selector: "The presentation of data and analyses in the prior section cannot settle questions of policy, nor decide the best balance of practices to meet the obligations and responsibilities of [the University of California] as an institution. A sober evaluation and interpretation of the data presented in this report, however, should inform policy decisions . . . and could help all stakeholders . . . understand better what the implications of various policy choices regarding the continued use of standardized tests for admission may be."[18]

In other words, data alone cannot drive the decision. The data presented cannot tell the university where to go, only what are "the implications of various policy choices," to use the words of the task force. This is what it looks like for statistics to be informative but not decisive for policy, as described in chapter 1. The lack of a straight path from data to decision partly reflects remaining uncertainties, such as how dropping the test would affect who applies to the university or how reweighting other admission factors would affect the diversity of admits. But resolving all these uncertainties would still not straighten the path from data to decision because of a diversity in how people prioritize among multiple goals. This feeds into whether people view something as large or small—some would look at the statistics presented in this chapter and the task force report and say that the effect of test scores on admission rates for high-GPA students is too large; others would call it small. What do you call it?

In recognizing the many dimensions of the issue, the task force warned against a formulaic path from data to decision: "A decision to make an admissions policy change at [the University of California] should not be driven solely by data and empirical findings. It should be driven by data together with a thoughtful evaluation of what purposes, goals, and mission are served well by the admissions process using the data it can gather."[19]

And so the task force submitted its report to university leadership in early 2020, several months after various groups sued the university for requiring that applicants submit standardized test scores.[20] In May 2021, the university settled the suit and agreed not to consider SAT or ACT scores in admission decisions through 2025.[21] In November of the same year, the university's Board of Regents decided to end the use of the tests indefinitely.[22]

Given the findings and recommendations of the task force, some might argue that university leadership ignored the evidence and caved to political pressure to the detriment of students.[23] Perhaps critique is warranted, but the policy aide—the task force in this instance—must remember that she is not the policy maker and is not accountable for the decision made. The Board of Regents and others with authority over the matter are responsible for the fruit of the policy change, whether good or bad, and this likely entered their minds as they weighed the issues.

Regardless of the decision, the policy aide should take satisfaction for having been right, that is, for having competently and transparently brought statistics to bear on the question at hand. The task force report did not shroud the issue with vague language and a barrage of uninterpreted statistics. Rather, it thoroughly examined key questions with insightful and transparent use of statistics that provided a clearer and deeper understanding of the issues. And of course, the policy is not settled. It never is because policy making is perpetual. Students that applied for enrollment in the fall of 2022 knew in advance that they did not have to take a standardized test. So, were admits more representative? Will they have lower GPAs in the coming years? Lower graduation rates?

Next Steps: How to Gauge the New Policy's Effect?

Imagine that you work in institutional research and academic planning in the Office of the President of the University of California several years after it ended its standardized test requirement. No studies have been published on the effects of the policy change, but university leadership wants to keep abreast of what the data are showing, especially regarding representation and academic performance. Which statistics would you calculate for what outcomes and samples? Which comparisons would you make, and what would you seek to learn from them? Think about it.

Start simple. One expected effect of dropping the test requirement is an increase in applications, especially from underrepresented groups. So consider comparing the total number of applications to the university and the composition of the applicant pool for 2020, 2021, and 2022. Did total applications increase? Did they increase for underrepresented students in absolute terms or as a share of applications?

How would we interpret differences from 2020 to 2021 and from 2021 to 2022? Students applying to enroll in 2020 or in 2021 had to take the tests (students applying for 2021 would have learned about the policy change after submitting their applications). Thus, the applicant pool for 2021 and prior years reflects the prior admission policy. Students applying for enrollment in 2022, in contrast, spent their senior year knowing that the university had dropped its standardized test requirement.

Comparing the applicant pool size and composition over 2020, 2021, and 2022, however, has a major limitation if we are using it to gain a sense of the initial effect of the policy change. The COVID-19 pandemic caused major disruptions that affected decisions about college. Most applications for fall 2020 would have been submitted before disruptions, but applications for fall 2021 would have been done in the midst of the pandemic. Moreover, the pandemic's effects on students varied, with students from low-income communities often suffering the greatest setback in learning.[24]

Recall from chapter 6 that good causal estimates from observational data use a change in policy over time, which we have, as well as variation in exposure to the policy change, which we lack if we compare the University of California only with itself over time. We need a control group that experienced the pandemic and other broad changes over time but that kept standardized test requirements. One control group might be private colleges and universities in California above a certain size and that kept the requirement; another could be public universities in other states. Creatively using such variation to estimate causal effects is the subject of books such as Angrist and Pischke's *Mastering 'Metrics*.[25]

Questions for Reflection and Discussion

1. Do you consider the influence of SAT scores on admission rates among high-GPA applicants (fig. 9.2, left side) large or small? Remember, anyone can pick an adjective. The work is in giving compelling reasons behind your pick.

2. The task force considered four measures of college success: first-year GPA, retention to second year, graduating on time, and GPA at graduation. Which is the most important (or weighty, to use the term from chapter 5) and why?

3. The task force measured how well SAT scores predict college success after accounting for high school GPA. Should it have accounted for additional variables, such as those capturing the socioeconomic background of the student? If so, which additional variables would you include and why?

4. A policy maker asks for a short memo on the effects of the University of California's decision to drop the use of standardized test scores in admission decisions. You don't have time to do original analysis, so you rely on the publicly available work of others:

   a. Does the evidence support a reinstatement of the test requirement?

   b. If yes, which one or two statistics is your answer most based on and why?

   c. If no, what statistical findings would convince you to reinstate it?

Acknowledgments

I am grateful to Richard DiSalvo, Bill Dunn, Max Harleman, Daniel Hurst, Grant Jacobsen, Don Kenkel, James Leslie, Casey Mulligan, Michael Poznansky, Neil Silveus, Jason Smith, Sara Melo de Weber, Garrett Weber, four anonymous reviewers, and the staff of the University of Chicago Press who contributed to sharpening the book. I also thank students at the University of Pittsburgh Graduate School of Public and International Affairs whose desire to understand the use of statistics in policy forced me to think harder and communicate more clearly about the topic. Much of the book also benefited from my colleagues at the White House Council of Economic Advisers, from whom I learned much. All remaining errors are mine.

Notes

Chapter One

1. The policy aide is therefore a broader role than the policy analyst described by Weimer and Vining, in their *Policy Analysis: Concepts and Practice* (Boston: Pearson, 2017), who serves a client by enumerating options and evaluating trade-offs.
2. Justine Hastings and Jesse M. Shapiro, "How Are SNAP Benefits Spent? Evidence from a Retail Panel," *American Economic Review* 108, no. 12 (2018): 3493–3540.
3. Eugene Bardach and Eric M. Patashnik, *A Practical Guide for Policy Analysis: The Eightfold Path to More Effective Problem Solving*, 6th ed. (Thousand Oaks, CA: Sage Publications and CQ Press, 2020).

Chapter Two

1. David S. Moore, George P. McCabe, and Bruce A. Craig, *Introduction to the Practice of Statistics* (New York: W. H. Freeman, Macmillan Learning, 2021).
2. Timothy W. Russell et al., "Estimating the Infection and Case Fatality Ratio for Coronavirus Disease (COVID-19) Using Age-Adjusted Data from the Outbreak on the Diamond Princess Cruise Ship, February 2020," *Eurosurveillance* 25, no. 12 (2020), https://doi.org/10.2807/1560-7917.ES.2020.25.12.2000256.
3. Russell et al., 25.
4. World Health Organization, "WHO Director-General's Opening Remarks at the Media Briefing on COVID-19—3 March 2020," https://www.who.int/director-general/speeches/detail/who-director-general-s-opening-remarks-at-the-media-briefing-on-covid-19---3-march-2020.

5. Russell et al., "Estimating the Infection and Case Fatality Ratio for Coronavirus," 25.

6. Xiao Chen et al., "Ratio of Asymptomatic COVID-19 Cases among Ascertained SARS-COV-2 Infections in Different Regions and Population Groups in 2020: A Systematic Review and Meta-Analysis Including 130 123 Infections from 241 Studies," *BMJ Open* 11, no. 12 (2021), https://doi.org/10.1136/bmjopen-2021-049752.

7. US Department of Agriculture Economic Research Service, "Principal Farm Operator Household Finances, 2018-22F," https://www.ers.usda.gov/data-products/farm-household-income-and-characteristics/.

8. US Department of Agriculture Economic Research Service, "Glossary," https://www.ers.usda.gov/topics/farm-economy/farm-household-well-being/glossary/.

9. Erik J. O'Donoghue et al., "Exploring Alternative Farm Definitions Implications for Agricultural Statistics and Program Eligibility" (Economic Information Bulletin No. 49, US Department of Agriculture Economic Research Service, Washington, DC, March 2009).

10. See Jeremy G. Weber and Dawn Marie Clay, "Who Does Dot Respond to the Agricultural Resource Management Survey and Does It Matter?," *American Journal of Agricultural Economics* 95, no. 3 (April 2013): table 3.

11. Weber and Clay, tables 5 and 6.

12. US Census Bureau, *Final Report of the Interagency Technical Working Group on Evaluating Alternative Measures of Poverty* (Washington, DC: US Census Bureau, 2021), https://www.census.gov/content/dam/Census/library/publications/2021/demo/EvaluatingAlternativeMeasuresofPoverty_08Jan2021.pdf.

13. Emily A. Shrider et al., *Income and Poverty in the United States: 2020* (Report No. P60-273) (Washington, DC: US Census Bureau, September 2021).

14. Phil Gramm and John F. Early, "The Truth about Income Inequality," *Wall Street Journal*, November 3, 2019, https://www.wsj.com/articles/the-truth-about-income-inequality-11572813786.

15. Michael Novogradac, "How Much Capital Have Opportunity Fund Managers Raised?," *Novogradac* (blog), June 28, 2019, https://www.novoco.com/notes-from-novogradac/how-much-capital-have-opportunity-fund-managers-raised.

16. Council of Economic Advisers, *The Impact of Opportunity Zones: An Initial Assessment* (Washington, DC: Council of Economic Advisers, August 2020).

Chapter Three

1. US Department of Commerce, "Publication of a Report on the Effect of Imports of Uranium on the National Security: An Investigation

Conducted under Section 232 of the Trade Expansion Act of 1962, as Amended," *Federal Register* 86, no. 145 (August 2, 2021): 41540–610.

2. US Energy Information Administration, *2018 Uranium Marketing Annual Report* (Washington, DC: US Energy Information Administration, May 2019); US Energy Information Administration, *2018 Domestic Uranium Production Report* (Washington, DC: US Energy Information Administration, May 2019).The weighted-average purchase price for 2010–2014 well exceeds $55 when adjusted for changes in the Producer Price Index for mining.

3. Catherine Lucey, Nomaan Merchant, and Jonathan Lemire, "Trump Threatens to Close Border with Mexico Next Week," *PBS NewsHour*, March 29, 2019, www.pbs.org/newshour/nation/trump-threatens-to -close-border-with-mexico-next-week.

4. Sam Meredith, "Oil Tanker Attacks in the Strait of Hormuz Requires an 'International Response,' US Envoy to Iran Says," CNBC, June 22, 2019, www.cnbc.com/2019/06/22/oil-tanker-attacks-in-the-strait-of -hormuz-requires-an-international-response-us-envoy-to-iran-says .html.

5. US Census Bureau, "American Community Survey Selected Economic Characteristics 2019: ACS 5-Year Estimates Data Profiles," https:// data.census.gov/cedsci/table?t=Income%20and%20Poverty&g= 1600000US5344725&y=2019&d=ACS%205-Year%20Estimates%20 Data%20Profiles&tid=ACSDP5Y2019.DP03.

6. Jeff Ernsthausen, Jesse Eisinger, and Paul Kiel, "The Secret IRS Files: Trove of Never-before-Seen Records Reveal How the Wealthiest Avoid Income Tax," ProPublica, June 8, 2021, https://www.propublica .org/article/the-secret-irs-files-trove-of-never-before-seen-records -reveal-how-the-wealthiest-avoid-income-tax.

7. Jeremy G. Weber, Yongsheng Wang, and Maxwell Chomas, "A Quantitative Description of State-Level Taxation of Oil and Gas Production in the Continental U.S," *Energy Policy* 96 (2016): 289–301.

8. US Census Bureau, "American Community Survey Selected Economic Characteristics 2019."

9. US Department of Agriculture Economic Research Service, "Principal Farm Operator Household Finances, 2018-22F."

10. Adam Looney, "Student Loan Forgiveness Is Regressive Whether Measured by Income, Education, or Wealth" (Working Paper No. 75, Hutchins Center on Fiscal and Monetary Policy at Brookings, January 2022).

11. The sample variance and standard deviation are descriptions of sample data and are distinct from sample-based estimates of the population variance and standard deviation. In one case you are describing a sample; in the other you are estimating a value that reflects the whole population after observing only part of it. Mathematically the difference is that when estimating the population variance, divide by $n - 1$ instead of n. The square root of this estimated population vari-

ance is then the estimated population standard deviation. We need the population standard deviation because the standard error seeks to describe the population of samples so to speak, which is the sampling distribution of the mean.

12. The term $n - 1$ cancels out of the numerator and denominator, so I have omitted it to simplify the expression.

13. Given an estimated slope coefficient of $\hat{\beta}$, the estimated intercept term is given by $\hat{\alpha} = \hat{\beta}\overline{PI} - \overline{CI}$. This works because the best-fit line passes through the mean of the data such that $\overline{CI} = \hat{\alpha} + \hat{\beta}\overline{PI}$.

Chapter Four

1. The modified equation is then

Road spending = −62 miles of road + 1.9 potholes

+ 1.2 household income × miles of road.

A reason for not using this equation for estimation is that it allows outlier neighborhoods with many miles of road to have an outsized effect on the coefficient estimates.

2. Council of Economic Advisers, *The State of Homelessness in America* (Washington, DC: Council of Economic Advisers, September 2019).

3. US Department of Housing and Urban Development, *Risk or Race: An Assessment of Subprime Lending Patterns in Nine Metropolitan Areas* (Washington, DC: US Department of Housing and Urban Development, August 2009).

4. See Carmen M. Gutierrez, "The Institutional Determinants of Health Insurance: Moving Away from Labor Market, Marriage, and Family Attachments under the ACA," *American Sociological Review* 83, no. 6 (2018): 1144–70; Grant D. Jacobsen, "An Examination of How Energy Efficiency Incentives Are Distributed across Income Groups," *Energy Journal* 40, no. 01 (January 2019): 171–98; Corbett Grainger and Andrew Schreiber, "Discrimination in Ambient Air Pollution Monitoring?," *AEA Papers and Proceedings* 109 (January 2019): 277–82.

Chapter Five

1. Robert P. Abelson, *Statistics as Principled Argument* (New York: Psychology Press, 2009).

2. J. Briskin and L. Yohannes, "Overview of US EPA's Study of the Potential Impacts of Hydraulic Fracturing for Oil and Gas on Drinking Water Resources," *ACS Symposium Series* (2015): 197–203.

3. US Environmental Protection Agency, "Assessment of the Potential Impacts of Hydraulic Fracturing for Oil and Gas on Drinking Water Resources," External Review Draft (June 2015).

4. The EPA reported spills and groundwater contamination separately for fracking chemical spills (151 spills with no groundwater contamination) and produced water spills (225 spills with eighteen cases of groundwater contamination).

5. Chris Mooney, Steven Mufson, and Brady Dennis, "EPA's Science Advisers Challenge Agency Report on the Safety of Fracking," *Washington Post*, October 27, 2021, https://www.washingtonpost.com /news/energy-environment/wp/2016/08/11/epas-science-advisers -challenge-agency-report-on-the-safety-of-fracking/.

6. US Environmental Protection Agency, *Hydraulic Fracturing for Oil and Gas: Impacts from the Hydraulic Fracturing Water Cycle on Drinking Water Resources in the United States* (Washington, DC: US Environmental Protection Agency, December 2016).

7. US Energy Information Administration, "Horizontally Drilled Wells Dominate U.S. Tight Formation Production," *Today in Energy*, June 6, 2019.

8. Stephen T. Ziliak and Deirdre McCloskey, "Lady Justice versus Cult of Statistical Significance," in *The Oxford Handbook of Professional Economic Ethics*, ed. George F. DeMartino and Deirdre McCloskey (New York: Oxford University Press, 2014), 351–64.

9. The numerator of the t-statistic is the difference in means (= 0.09 = 1.09 – 1.0); the denominator is the square root of the sum of the group variances divided by their group size ($\sqrt{(2.250/1000) + (1/1000)}$). The ratio of the two is then 1.58 (= 0.09 / 0.057).

10. Jeremy G. Weber and Dawn Marie Clay, "Who Does Dot Respond to the Agricultural Resource Management Survey and Does It Matter?," *American Journal of Agricultural Economics* 95, no. 3 (April 2013): 755–71.

11. Jacob Cohen, "A Power Primer," *Psychological Bulletin* 112, no. 1 (1992): 155–59.

12. Henian Chen et al., "How Big Is a Big Odds Ratio? Interpreting the Magnitudes of Odds Ratios in Epidemiological Studies," *Communications in Statistics—Simulation and Computation* 39, no. 4 (2010): 860-64.

13. Brady W. Allred et al., "Ecosystem Services Lost to Oil and Gas in North America," *Science* 348, no. 6233 (2015): 401–2.

14. Benefit-cost ratios are a type of effect-to-effort ratio where all effects and efforts are collapsed into one unit (dollars). Assigning a dollar value to everything makes it possible to compare across all effects and efforts but requires assumptions about the substitutability of one good for another such as good health for cheap electricity or biodiversity for cheap food.

Chapter Six

1. Justin Peters, "When Ice Cream Sales Rise, so Do Homicides: Coincidence, or Will Your Next Cone Murder You?," *Slate*, July 9,

2013, https://slate.com/news-and-politics/2013/07/warm-weather
-homicide-rates-when-ice-cream-sales-rise-homicides-rise
-coincidence.html.

2. Joshua D. Angrist and Jörn-Steffen Pischke, "The Credibility Revolution in Empirical Economics: How Better Research Design Is Taking
the Con out of Econometrics," *Journal of Economic Perspectives* 24, no. 2
(2010): 3–30.

3. C. Kirabo Jackson, "Does School Spending Matter? The New Literature on an Old Question" (Working Paper No. w25368, National
Bureau of Economic Research, Cambridge, MA, December 2018).

4. The 2017 and 2019 reports *Income and Poverty in the United States*
indicate 39.7 million people in poverty in 2017 compared to 34.0 million in 2019. See Fontenot, Kayla, Jessica Semega, and Melissa Kollar,
Income and Poverty in the United States: 2017 (Washington, DC: US
Census Bureau, September 2018), https://www.census.gov/content
/dam/Census/library/publications/2018/demo/p60-263.pdf; Jessica
Semega et al., *Income and Poverty in the United States: 2019* (Washington, DC: US Census Bureau, September 2020).

5. Economic Innovation Group, "Facts & Figures," https://eig.org
/opportunity-zones/facts-figures/.

6. Novogradac, "Opportunity Funds List," https://www.novoco.com
/resource-centers/opportunity-zone-resource-center/opportunity
-funds-listing.

7. Sarah C. Cambon and Danny Dougherty, "States That Cut Unemployment Benefits Saw Limited Impact on Job Growth," *Wall Street
Journal*, September 1, 2021, https://www.wsj.com/articles/states-that
-cut-unemployment-benefits-saw-limited-impact-on-job-growth
-11630488601.

8. Harry J. Holzer et al., "Did Pandemic Unemployment Benefits Reduce
Employment? Evidence from Early State-Level Expirations in June
2021" (Working Paper No. w29575, National Bureau of Economic
Research, Cambridge, MA, December 2021).

9. Holzer et al., "Did Pandemic Unemployment Benefits Reduce
Employment?"

10. Sarah C. Cambon and Danny Dougherty, "States That Cut Unemployment Benefits Saw Limited Impact on Job Growth," *Wall Street
Journal*, September 1, 2021, https://www.wsj.com/articles/states-that
-cut-unemployment-benefits-saw-limited-impact-on-job-growth
-11630488601.

11. Assuming that the unemployment rates are calculated from data on
the entire labor force, it might seem like the sampling error for the
difference in mean unemployment rates should be 0. But consider
all the random background events happening in mid-2020 and that
potentially affected unemployment in some states and not others. The
unemployment rate data for 2020 therefore reflect a particular state of
the world, one of many potential draws of the state of the world, each

with its own random mix of events. A different draw for the state of the world could yield different estimates of causal effects. Thus, the standard error of the estimated causal effect is still meaningful even if the underlying unemployment rates are based on a census of the labor force.

12. Angus Deaton, "Randomization in the Tropics Revisited: A Theme and Eleven Variations" (Working Paper No. w27600, National Bureau of Economic Research, Cambridge, MA, July 2020).

13. Jean Drèze, "Evidence, Policy, and Politics," *Ideas for India*, https://www.ideasforindia.in/topics/miscellany/evidence-policy-and-politics.html.

Chapter Seven

1. Deirdre N. McCloskey, *Economical Writing* (Prospect Heights, IL: Waveland Press, 2000), 12.

2. Pittsburgh Water and Sewer Authority, *Important Information about Lead in Your Drinking Water—Pittsburgh* (Pittsburgh: Pittsburgh Water and Sewer Authority, July 2016), https://apps.pittsburghpa.gov/pwsa/DEP-PWSA-PN-7-25-16.pdf. The results are reported in ranges, but I use the midpoint to approximate the full distribution and calculate the mean.

3. US Environmental Protection Agency, "Lead and Copper Rule," https://www.epa.gov/dwreginfo/lead-and-copper-rule. https://www.epa.gov/dwreginfo/lead-and-copper-rule.

4. To keep the details in the figure environment, the figure note could state that appreciation is based off an estimated median housing price for one-unit noncondominium properties produced by the Federal Housing Finance Agency (FHFA).

Chapter Eight

1. Eugene Bardach and Eric M. Patashnik, in *A Practical Guide for Policy Analysis The Eightfold Path to More Effective Problem Solving* (Thousand Oaks, CA: Sage and CQ Press, 2020), 16.

2. Neil M. Ferguson et al., *Report 9: Impact of Non-Pharmaceutical Interventions (NPIs) to Reduce COVID-19 Mortality and Healthcare Demand* (London: Imperial College COVID-19 Response Team, March 16, 2020).

3. The briefing can be seen on YouTube at "Members of the Coronavirus Task Force Hold a Briefing," March 13, 2020, posted by US Department of State, https://www.youtube.com/watch?v=e9v8ZZd1P0M.

4. Mark Landler and Stephen Castle, "Behind the Virus Report That Jarred the US and the UK to Action," *New York Times*, March 17, 2020,

www.nytimes.com/2020/03/17/world/europe/coronavirus-imperial
-college-johnson.html.

5. Austan Goolsbee and Chad Syverson, "Fear, Lockdown, and Diversion: Comparing Drivers of Pandemic Economic Decline 2020," *Journal of Public Economics* 193 (2021): 104311.

6. Observational studies are those in which the variation in the causal variable is from the normal churn of life, not from purposeful manipulation by researchers.

7. Maggie Fox and Stacey Naggiar, "Rising Levels of Toxic Gas Found in Homes near Fracking Sites," NBCNews.com, April 9, 2015, www.nbcnews.com/health/health-news/could-fracking-raise-lung-cancer-risk-n338146.

8. Katie Jo Black et al., "Fracking and Indoor Radon: Spurious Correlation or Cause for Concern?," *Journal of Environmental Economics and Management* 96 (2019): 255–73.

9. John P. A. Ioannidis, "A Fiasco in the Making? As the Coronavirus Pandemic Takes Hold, We Are Making Decisions without Reliable Data," *STAT* (March 17, 2020): 1–6. Ioannidis used the *Diamond Princess* data and adjusted it for the age structure of the US population.

10. John P. Ioannidis, "Reconciling Estimates of Global Spread and Infection Fatality Rates of COVID-19: An Overview of Systematic Evaluations," *European Journal of Clinical Investigation* 51, no. 5 (September 2021): https://doi.org/10.1111/eci.13554.

Chapter Nine

1. University of California Academic Senate, *Report of the UC Academic Council Standardized Testing Task Force (STTF)* (STFF Report) (Oakland: University of California, January 2020), 3, https://senate.universityofcalifornia.edu/_files/committees/sttf/sttf-report.pdf.

2. STTF Report, 4 and 65.

3. STTF Report, 10. The report is quoting University Policy 2102.

4. STTF Report, introductory letter by the report cochairs.

5. STTF Report, introductory letter by the report cochairs.

6. STTF Report, 77.

7. STTF Report, 68.

8. STTF Report, 20.

9. The intuition behind this can been gained by considering a university to which 99 percent of applicants are admitted. Even if the test scores of admitted students are highly correlated with graduation rates, the practical relevance of the correlation is undone because most students are admitted anyway, so the average score of applicants and admits is similar. The opposite is true when the admission rate is low, in which case changing admission rates based on test scores will widen the

difference in scores among applicants and admits and, by extension, graduation rates.

10. STTF Report, 57.
11. STTF Report, 22.
12. STTF Report, 26–28.
13. STTF Report, 16.
14. The STTF Report's table 3C-4 on page 43 has exhaustive breakouts by race and ethnicity and also by first-generation status.
15. STTF Report, 31.
16. STTF Report, 48.
17. STTF Report, 69.
18. STTF Report, 63.
19. STTF Report, 68–69.
20. Larry Gordon, "Lawsuits Seek to End University of California's SAT or ACT Test Requirement for Freshman Admission," *EdSource*, December 10, 2019.
21. Giulia McDonnell Nieto del Rio, "University of California Will No Longer Consider SAT and ACT Scores," *New York Times*, May 15, 2021.
22. Michael T. Nietzel, "University of California Reaches Final Decision: No More Standardized Admission Testing," *Forbes*, November 19, 2021.
23. Caitlin Flanagan, "The University of California Is Lying to Us," *The Atlantic*, July 22, 2021.
24. Bianca Vasquez Toness, Sharon Lurye, and Howard Blume, "Massive Learning Setbacks Show COVID's Sweeping Toll on Kids across the Country, Data Show," *Los Angeles Times*, October 28, 2022.
25. Joshua D. Angrist and Jörn-Steffen Pischke, *Mastering 'Metrics: The Path from Cause to Effect* (Princeton, NJ: Princeton University Press, 2014).

Bibliography

Abelson, Robert P. *Statistics as Principled Argument.* New York: Psychology Press, 2009.

Allred, Brady W., W. Kolby Smith, Dirac Twidwell, Julia H. Haggerty, Steven W. Running, David E. Naugle, and Samuel D. Fuhlendorf. "Ecosystem Services Lost to Oil and Gas in North America." *Science* 348, no. 6233 (2015): 401–2.

Angrist, Joshua D., and Jörn-Steffen Pischke. "The Credibility Revolution in Empirical Economics: How Better Research Design Is Taking the Con Out of Econometrics." *Journal of Economic Perspectives* 24, no. 2 (2010): 3–30.

———. *Mastering 'Metrics: The Path from Cause to Effect.* Princeton, NJ: Princeton University Press, 2014.

Bardach, Eugene, and Eric M. Patashnik. *A Practical Guide for Policy Analysis: The Eightfold Path to More Effective Problem Solving.* 6th ed. Thousand Oaks, CA: Sage Publications and CQ Press, 2020.

Black, Katie Jo, Shawn J. McCoy, and Jeremy G. Weber. "Fracking and Indoor Radon: Spurious Correlation or Cause for Concern?" *Journal of Environmental Economics and Management* 96 (2019): 255–73.

Briskin, J., and L. Yohannes. "Overview of U.S. EPA's Study of the Potential Impacts of Hydraulic Fracturing for Oil and Gas on Drinking Water Resources." *ACS Symposium Series* (2015): 197–203.

Cambon, Sarah C., and Danny Dougherty. "States That Cut Unemployment Benefits Saw Limited Impact on Job Growth." *Wall Street Journal*, September 1, 2021. https://www.wsj.com/articles/states-that-cut-unemployment -benefits-saw-limited-impact-on-job-growth-11630488601.

Chen, Henian, Patricia Cohen, and Sophie Chen. "How Big Is a Big Odds Ratio? Interpreting the Magnitudes of Odds Ratios in Epidemiological Studies." *Communications in Statistics—Simulation and Computation* 39, no. 4 (2010): 860–64.

Chen, Xiao, Ziyue Huang, Jingxuan Wang, Shi Zhao, Martin Chi-Sang Wong, Ka Chun Chong, Daihai He, and Jinhui Li. "Ratio of Asymptomatic COVID-19 Cases among Ascertained SARS-COV-2 Infections in Different Regions and Population Groups in 2020: A Systematic Review and Meta-Analysis Including 130 123 Infections from 241 Studies." *BMJ Open* 11, no. 12 (2021). https://doi.org/10.1136/bmjopen-2021-049752.

Cohen, Jacob. "A Power Primer." *Psychological Bulletin* 112, no. 1 (1992): 155–59.

Council of Economic Advisers. *The Impact of Opportunity Zones: An Initial Assessment.* Washington, DC: Council on Economic Advisers, August 2020.

———. *The State of Homelessness in America.* Washington, DC: Council on Economic Advisers, September 2019.

Deaton, Agnus. "Randomization in the Tropics Revisited: A Theme and Eleven Variations." Working Paper No. w27600, National Bureau of Economic Research, Cambridge, MA, July 2020.

Drèze, Jean. "Evidence, Policy, and Politics." *Ideas for India.* https://www.ideasforindia.in/topics/miscellany/evidence-policy-and-politics.html.

Economic Innovation Group. "Facts & Figures." https://eig.org/opportunity-zones/facts-figures/.

Ernsthausen, Jeff, and Jesse Eisinger. "The Secret IRS Files: Trove of Never-before-Seen Records Reveal How the Wealthiest Avoid Income Tax." ProPublica, June 8, 2021. https://www.propublica.org/article/the-secret-irs-files-trove-of-never-before-seen-records-reveal-how-the-wealthiest-avoid-income-tax.

Ferguson, Neil M., Daniel Layton, Gemma Nedjati-Gilani, Natsuko Imai, and Kylie Ainslie. *Report 9: Impact of Non-Pharmaceutical Interventions (NPIs) to Reduce COVID-19 Mortality and Healthcare Demand.* London: Imperial College COVID-19 Response Team, March 16, 2020.

Fontenot, Kayla, Jessica Semega, and Melissa Kollar. *Income and Poverty in the United States: 2017.* Washington, DC: US Census Bureau, September 2018. https://www.census.gov/content/dam/Census/library/publications/2018/demo/p60-263.pdf.

Fox, Maggie, and Stacey Naggiar. "Rising Levels of Toxic Gas Found in

Homes near Fracking Sites." NBCNews.com, April 9, 2015, https://www
.nbcnews.com/health/health-news/could-fracking-raise-lung-cancer
-risk-n338146.

Goolsbee, Austan, and Chad Syverson. "Fear, Lockdown, and Diversion:
Comparing Drivers of Pandemic Economic Decline 2020." *Journal of
Public Economics* 193 (2021): 104311.

Gordon, Larry. "Lawsuits Seek to End University of California's SAT or ACT
Test Requirement for Freshman Admission." *EdSource*, December 10,
2019.

Grainger, Corbett, and Andrew Schreiber. "Discrimination in Ambient Air
Pollution Monitoring?" *AEA Papers and Proceedings* 109 (2019): 277–82.
https://doi.org/10.1257/pandp.20191063.

Gramm, Phil, and John F. Early. "The Truth about Income Inequality." *Wall
Street Journal*, November 3, 2019. https://www.wsj.com/articles/the-truth
-about-income-inequality-11572813786.

Gutierrez, Carmen M. "The Institutional Determinants of Health Insurance:
Moving Away from Labor Market, Marriage, and Family Attachments
under the ACA." *American Sociological Review* 83, no. 6 (2018): 1144–70.

Hastings, Justine, and Jesse M. Shapiro. "How Are SNAP Benefits Spent? Evidence from a Retail Panel." *American Economic Review* 108, no. 12 (2018):
3493–3540.

Holzer, Harry J., R. Glenn Hubbard, and Michael R. Strain. "Did Pandemic
Unemployment Benefits Reduce Employment? Evidence from Early
State-Level Expirations in June 2021." Working Paper No. w29575, Na-
tional Bureau of Economic Research, Cambridge, MA, December 2021.

Ioannidis, John P. A. "A Fiasco in the Making? As the Coronavirus Pandemic
Takes Hold, We Are Making Decisions without Reliable Data." *STAT*
(March 17, 2020): 1–6.

———. "Reconciling Estimates of Global Spread and Infection Fatality Rates
of COVID-19: An Overview of Systematic Evaluations." *European Journal
of Clinical Investigation* 51, no. 5 (2021). https://doi.org/10.1111/eci.13554.

Jackson, C. Kirabo. "Does School Spending Matter? The New Literature on an
Old Question." Working Paper No. w25368, National Bureau of Economic
Research, Cambridge, MA, December 2018.

Jacobsen, Grant D. "An Examination of How Energy Efficiency Incentives
Are Distributed across Income Groups." *Energy Journal* 40, no. 1 (2019):
171–98.

Landler, Mark, and Stephen Castle. "Behind the Virus Report That Jarred the

US and the UK to Action." *New York Times*, March 17, 2020, https://www
.nytimes.com/2020/03/17/world/europe/coronavirus-imperial-college
-johnson.html.

Looney, Adam. "Student Loan Forgiveness Is Regressive Whether Measured
by Income, Education, or Wealth." Working Paper No. 75, Hutchins Center
on Fiscal and Monetary Policy at Brookings, January 2022.

Lucey, Catherine, Nomaan Merchant, and Jonathan Lemire. "Trump Threatens to Close Border with Mexico Next Week." *PBS NewsHour*, March 29,
2019, https://www.pbs.org/newshour/nation/trump-threatens-to-close
-border-with-mexico-next-week.

McCloskey, Deirdre N. *Economical Writing*. Prospect Heights, IL: Waveland
Press, 2000.

Meredith, Sam. "Oil Tanker Attacks in the Strait of Hormuz Requires an
'International Response,' US Envoy to Iran Says." CNBC, June 22, 2019,
www.cnbc.com/2019/06/22/oil-tanker-attacks-in-the-strait-of-hormuz
-requires-an-international-response-us-envoy-to-iran-says.html.

Mooney, Chris, Steven Mufson, and Brady Dennis. "EPA's Science Advisers Challenge Agency Report on the Safety of Fracking." *Washington
Post*, October 27, 2021, https://www.washingtonpost.com/news/energy
-environment/wp/2016/08/11/epas-science-advisers-challenge-agency
-report-on-the-safety-of-fracking/.

Moore, David S., George P. McCabe, and Bruce A. Craig. *Introduction to the
Practice of Statistics*. New York: W. H. Freeman, Macmillan Learning, 2021.

Nieto del Rio, Giulia McDonnell. "University of California Will No Longer
Consider SAT and ACT Scores." *New York Times*, May 15, 2021.

Nietzel, Michael, T. "University of California Reaches Final Decision: No
More Standardized Admission Testing." *Forbes*, November 19, 2021.

Novogradac. "Opportunity Funds List." https://www.novoco.com/resource
-centers/opportunity-zone-resource-center/opportunity-funds-listing.

Novogradac, Michael. "How Much Capital Have Opportunity Fund Managers Raised?" *Novogradac* (blog), June 28, 2019, https://www.novoco.com
/notes-from-novogradac/how-much-capital-have-opportunity-fund
-managers-raised.

O'Donoghue, Erik J., Penni Korb, David E. Banker, and Robert A Hoppe.
"Exploring Alternative Farm Definitions Implications for Agricultural Sta-
tistics and Program Eligibility." Economic Information Bulletin No. 49,
USDA Economic Research Service, Washington, DC, March 2009.

Peters, Justin. "When Ice Cream Sales Rise, so Do Homicides: Coincidence, or Will Your Next Cone Murder You?" *Slate*, July 9, 2013, https://slate.com /news-and-politics/2013/07/warm-weather-homicide-rates-when-ice -cream-sales-rise-homicides-rise-coincidence.html.

Pittsburgh Water and Sewer Authority. *Important Information about Lead in Your Drinking Water—Pittsburgh*. Pittsburgh: Pittsburgh Water and Sewer Authority, July 2016. https://apps.pittsburghpa.gov/pwsa/DEP-PWSA-PN -7-25-16.pdf.

Russell, Timothy W., Joel Hellewell, Christopher I. Jarvis, Kevin van Zand-voort, Sam Abbott, Ruwan Ratnayake, Stefan Flasche, Rosalind M. Eggo, W. John Edmunds, and Adam J. Kucharski. "Estimating the Infection and Case Fatality Ratio for Coronavirus Disease (COVID-19) Using Age-Adjusted Data from the Outbreak on the Diamond Princess Cruise Ship, February 2020." *Eurosurveillance* 25, no. 12 (2020). https://doi.org/10.2807 /1560-7917.ES.2020.25.12.2000256.

Semega, Jessica, Melissa Kollar, Emily A. Shrider, and John F. Creamer. *Income and Poverty in the United States: 2019*. Washington, DC: US Census Bureau, September 2020.

Shrider, Emily A., Melissa Kollar, Frances Chan, and Jessica Semega. *Income and Poverty in the United States: 2020*. Report No. P60-273. Washington, DC: US Census Bureau, September 2021.

University of California Academic Senate. *Report of the UC Academic Council Standardized Testing Task Force (STTF)*. Oakland: University of California, January 2020, https://senate.universityofcalifornia.edu/_files /committees/sttf/sttf-report.pdf.

US Census Bureau. "American Community Survey Selected Economic Characteristics 2019: ACS 5-Year Estimates Data Profiles." https:// data.census.gov/cedsci/table?t=Income%20and%20Poverty&g= 1600000US5344725&y=2019&d=ACS%205-Year%20Estimates%20Data %20Profiles&tid=ACSDP5Y2019.DP03.

———. *Final Report of the Interagency Technical Working Group on Evaluating Alternative Measures of Poverty*. Washington, DC: US Census Bureau, 2021. https://www.census.gov/content/dam/Census/library/publications /2021/demo/p60-274.pdf.

US Department of Agriculture Economic Research Service. "Glossary." https://www.ers.usda.gov/topics/farm-economy/farm-household-well -being/glossary/.

———. "Principal Farm Operator Household Finances, 2018-22F." https:// www.ers.usda.gov/data-products/farm-household-income-and -characteristics/.

US Department of Commerce. "Publication of a Report on the Effect of Imports of Uranium on the National Security: An Investigation Conducted Under Section 232 of the Trade Expansion Act of 1962, as Amended." *Federal Register* 86, no. 145 (August 2, 2021): 41540–610.

US Department of Housing and Urban Development. *Risk or Race: An Assessment of Subprime Lending Patterns in Nine Metropolitan Areas.* Washington, DC: US Department of Housing and Urban Development, August 2009.

US Energy Information Administration. *2018 Domestic Uranium Production Report.* Washington, DC: US Energy Information Administration, May 2019.

———. *2018 Uranium Marketing Annual Report.* Washington, DC: US Energy Information Administration, May 2019.

———. "Horizontally Drilled Wells Dominate US Tight Formation Production." June 6, 2019.

US Environmental Protection Agency. "Assessment of the Potential Impacts of Hydraulic Fracturing for Oil and Gas on Drinking Water Resources." External Review Draft, June 2015.

———. *Hydraulic Fracturing for Oil and Gas: Impacts from the Hydraulic Fracturing Water Cycle on Drinking Water Resources in the United States.* Washington, DC: US Environmental Protection Agency, December 2016.

———. "Lead and Copper Rule." https://www.epa.gov/dwreginfo/lead-and -copper-rule.

Vasquez Toness, Bianca, Sharon Lurye, and Howard Blume. "Massive Learning Setbacks Show Covid's Sweeping Toll on Kids across the Country, Data Show." *Los Angeles Times*, October 28, 2022.

Weber, Jeremy G., and Dawn Marie Clay. "Who Does Dot Respond to the Agricultural Resource Management Survey and Does It Matter?" *American Journal of Agricultural Economics* 95, no. 3 (2013): 755–71.

Weber, Jeremy G., Yongsheng Wang, and Maxwell Chomas. "A Quantitative Description of State-Level Taxation of Oil and Gas Production in the Continental U.S." *Energy Policy* 96 (2016): 289–301.

Weimer, David Leo, and Aidan R. Vining. *Policy Analysis: Concepts and Practice.* Boston: Pearson, 2017.

World Health Organization. "WHO Director-General's Opening Remarks at the Media Briefing on COVID-19—3 March 2020." https://www.who

.int/director-general/speeches/detail/who-director-general-s-opening
-remarks-at-the-media-briefing-on-covid-19---3-march-2020.

Ziliak, Stephen T., and Deirdre McCloskey. "Lady Justice versus Cult of Sta-
tistical Significance: Oomph-less Science and the New Rule of Law." In
The Oxford Handbook of Professional Economic Ethics, edited by George F.
DeMartino and Deirdre McCloskey, 351–64. New York: Oxford University
Press, 2014.

Index

Page numbers in italics refer to figures and tables.

Printed and bound by CPI Group (UK) Ltd, Croydon, CR0 4YY

11/08/2024

14539262-0001